W9-AQE-437

WATCHING THE WORLD'S WEATHER

WILLIAM JAMES BURROUGHS

WATCHING THE WORLD'S WEATHER

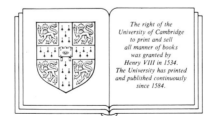

The right of the
University of Cambridge
to print and sell
all manner of books
was granted by
Henry VIII in 1534.
The University has printed
and published continuously
since 1584.

CAMBRIDGE UNIVERSITY PRESS

CAMBRIDGE

NEW YORK   PORT CHESTER

MELBOURNE   SYDNEY

ST. PHILIP'S COLLEGE LIBRARY

Published by the Press Syndicate of the University of Cambridge
The Pitt Building, Trumpington Street, Cambridge CB2 1RP
40 West 20th Street, New York NY 10011, USA
10 Stamford Road, Oakleigh, Melbourne 3166, Australia

© Cambridge University Press 1991

First published 1991

Printed by BAS Printers Limited,
Over Wallop, Hampshire

British Library cataloguing in publication data

Burroughs, William J. (William James)
Watching the world's weather.
1. Weather
I. Title
551.5

Library of Congress cataloguing in publication data

Burroughs, William James.
Watching the world's weather / William James Burroughs.
p.   cm.
Includes bibliographical references.
ISBN 0-521-34342-9
1. Astronautics in meteorology. 2. Meteorological satellites.
I. Title
QC879.5.B87   1990
551.6′3—dc20   90-43348   CIP

ISBN 0 521 34342 9 hardback

Contents

Contents

Contents

Preface

TO ANYONE with an interest in meteorology, the images obtained from satellites have provided a new perspective on the weather. For those who over the years had learnt to read the local conditions and combine these with the wider picture provided in weather maps the view from space opened up an added dimension. To those less well versed in meteorological theory they provided a new if sometimes perplexing insight into the endless variety of the weather. With the coming of the space age the whole swirling mass of the atmosphere with its patterns, complex three-dimensional motions, and recognisable features was laid out for all to see, but the translation of the visually appealing images into practical information has involved a lengthy process of technical development and emerging physical understanding. This book considers how the weather satellite has become an integral part of modern meteorology and how routine observations from space have been used not only by weathermen but also by others interested in a range of Earth sciences.

The practical relevance of weather satellites to other Earth sciences provides a second reason for this book. At this time of growing environmental concern about the potential impact of human activities on the climate, it is important to have a balanced picture of what is happening on a global scale. All too often local changes or short-term events are cited as evidence of a global trend. It is essential that we are not misled by such selective interpretation. Weather satellites offer the only means of obtaining reliable measurements of changes in such climatically important features as cloudiness, sea surface temperatures, and the extent of sea ice. After 30 years of satellite meteorology we are only beginning to see the quality of observations which can be used to make reliable conclusions to be drawn about changes in the global climate. In this context the detection of the 'ozone hole' over Antarctica showed how important it is to have the combination of ground-based observations which can verify otherwise suspect satellite measurements. So, now is a good time to take stock of how satellite meteorology has developed and to consider the central role it will play in the coming years in assessing whether the Earth's climate is changing.

Acknowledgements

IN PRODUCING this book I am indebted to a large number of people who have provided helpful information and, in particular, have been instrumental in tracking down many of the images presented here. These include, in alphabetical order: P. E. Baylis, C. C. Bradley, G. C. Bridge, R. J. H. Brush, J. A. Coakley, Jr, P. Cornillon, M. A. Geller, T. Gregory, J. E. Harries, K. Herrington, R. Hide, J. F. Le Marshall, M. Matson, J. R. Milford, V. Ramanathan, W. R. Rose, Jr, R. W. Saunders, M. Schroeberl, T. M. L. Wigley, W. S. Wilson, K. Woodley, J. A. Woods and H. W. Yates. In addition, the following organisations have been particularly helpful: the NASA Oceanic Processes Branch, the UK Meteorological Office Visual Aids Unit, The University of Dundee Department of Electrical Engineering and Electronics and the University of Wisconsin Space Science and Engineering Center.

Finally, I have to thank my wife for shouldering the burden of turning my much altered manuscript into typed output fit for the publishers and for dealing with the frequent amendments that were needed along the way.

1

Introduction

*'He that observeth the wind shall not sow;
and he that regardeth the clouds shall not reap.'*
Ecclesiastes 11.4

OVER THE CENTURIES man has tried to unravel the complexities of the weather. From the earliest times the continual changing of the elements has fascinated earth-bound viewers. But for most of history this fascination could only be exercised from a local perspective and enshrined in folklore. With the development of the science of meteorology in the last two centuries a wider picture of how the global atmosphere operates has gradually been built up. While much of the functioning of the atmosphere has been inferred from the Earth's surface and supplemented by measurements from balloons and aircraft, it is only with the advent of satellite technology that a truly global view of the weather has begun to emerge. This book describes the development of satellite technology and the effect it has had on our understanding of the weather.

1.1 The value of a view from space

The one thing about the weather that is certain is that it is always changing. These changes occur over a wide range of physical dimensions and time scales. At the local level sudden thunderstorms, squalls, or tornadoes occur in many parts of the world causing loss of life and damage to agriculture and property. These events are often associated with larger-scale weather systems which are easily seen from space. On a grander scale hurricanes and cyclones in the tropics and intense depressions at higher latitudes are instantly recognisable from space. For centuries such storms caused widespread damage and loss of life, especially in coastal communities. Extreme examples include the hurricane that hit Galveston, Texas in 1900 killing over 6000 people; the cyclone that swept in from the Bay of Bengal, Bangladesh

Fig. 1.1. Viewed from a satellite over the equator much of the Earth's weather can be seen in a single image (with permission of Japanese Meteorological Agency).

in 1970 killing over a quarter of a million people; and, the storm that swept down the North Sea in 1953 flooding low-lying areas of eastern England and the Nether-lands and leaving over 2000 dead. The advent of weather satellites has dramatically improved forecasts of the movement of such storms and sharply reduced the loss of life. This is in spite of the fact that, what appears to be increasingly aberrant weather, has led to mounting damage to property.

On a longer time scale weather-related events such as the crop failure in India in 1972 or the frequent poor harvests in the Soviet Union in the 1970s, and the fuel crises in Britain in February 1947 and in the eastern United States in January

Fig. 1.2. An example of the damage caused when severe gales lashed the British Isles on 25 January 1990, killing 47 people in Britain and a further 48 across northern Europe.

1977, illustrate how spells of abnormal weather can have a profound impact on national economies. While, on a still longer time scale the frequent droughts in sub-Saharan Africa (the Sahel) since the late 1960s have both touched the conscience of the Western world and raised worrying questions about possible permanent changes in the global climate.

These more lengthy aberrations are but a tiny sample of the continual changes that are an essential feature of the global weather system. If we are to understand the atmospheric circulation patterns which lead to bouts of extreme weather in different parts of the world, we must have a complete picture of how the various components fit together. This involves measuring the behaviour of the atmosphere and the oceans, changes in the vegetation, moisture and snow cover on land, and the extent of polar pack ice. Only through a thorough understanding of the interaction of these features will it be possible to produce useful forecasts over periods of months or even years ahead. Weather satellites have already made major contributions to advances in recent years and will be essential for future progress.

ST. PHILIP'S COLLEGE LIBRARY

1.2 The emerging global view

Satellite observations are now part of everyday life. On television and in the news-papers, images and measurements from space are frequently used to produce pub-lished weather forecasts. But the development of satellite technology has been a lengthy process which has involved the development of an entire range of new technologies including increasingly sophisticated instruments to measure various features of the atmosphere and the surface below, and ever greater demands on data transmission, assimilation techniques, and computer analysis. Only by advan-cing on a broad technological front has it been possible for weather satellites to become such an essential part of modern meteorology. The principles underlying these technological areas and the practical approaches adopted for each technique will be explored in this book.

The other area which has required sustained effort is the building up of a set of measurements of global parameters which previously had not been observed. Measurement of the real extent of arctic and antarctic pack ice, and continental snow cover in the Northern Hemisphere, and the seasonal variation in global cloudi-ness are examples of climatically important factors which, prior to the advent of satellites, had never been quantified. Now, after years of observations, the fluctua-tions in such parameters may represent the key to understanding longer-term changes in our weather.

This understanding is of particular importance in reaching an early conclusion about whether the observed warming trend in the global climate is due to the 'Green-house Effect'. Many scientists see the extremes of the 1980s as being the first signs of the impact of manmade pollution which is altering the Earth's climate. But both the accurate monitoring of global changes in the climate and shifts in the weather patterns depend on better measurements. Without these there is a danger of jumping to hasty and ill-advised conclusions. What is needed is not only improved obser-vations about how various components of the climatic system vary from year to year but also basic measurements of how some of the components control the climate.

Nowhere is this need for better measurements more urgent than in studying the role of clouds. The current computer models which are used to predict the possible scale of the Greenhouse Effect exhibit major differences of the treatment of clouds. Without reliable figures of just how much effect changes in global cloudiness could have on the climate, it will be impossible to produce trustworthy estimates on how the climate is likely to change as greenhouse gases build up in the atmosphere. In recent years improved satellite observations have started to produce results of the quality needed to tackle this problem. So future satellite studies hold the key both to the reliable measurements of how the climate behaves and just how much it is changing. These studies offer the prospect of improving the input to the computer models which are needed to make predictions about how these changes may develop in the future.

A more immediate benefit of satellite observation has been in the shorter-term forecasts. Rapid interpretation of satellite pictures has developed into a fine art. Observations from geostationary satellites have been combined with ground-based radar systems which have led to much improved forecasts of rainfall over a period of up to 12–24 hours ahead. In the slightly longer term, satellite observations are now an integral part of weather forecasts 3–7 days in advance. It is estimated that without measurements from space, the quality of these forecasts would be degraded by at least a day, so that a ground-based prediction 4 days ahead would be no better than one 5 days ahead employing available satellite data.

The story of the development of satellite meteorology can be presented in terms of the principal components; first, how satellite technology has been developed to observe the different facets of the global weather system; second, how the products of the technology have been used to build up and improve our understanding of the functioning of the Earth's weather; third, the exploitation of both the data and the understanding to improve weather forecasting to provide real benefits; and finally, how work up to the present has exposed gaps in our knowledge and what developments are planned to fill them.

2

The global weather machine

'So, naturalists observe, a flea
Hath smaller fleas that on him prey;
And these have smaller fleas to bite 'em,
And so proceed ad infinitum.'
Jonathan Swift

IF WE ARE TO UNDERSTAND the technology of weather satellites, we must first consider the essential elements of the global weather machine. Only by understanding the various physical processes governing the behaviour of the atmosphere is it possible to appreciate what can be seen from space. In doing this the most important features are the way in which the Earth remains in radiative balance with the Sun, and how atmospheric motions transport energy over the face of the Earth. This chapter will examine the principles of these two essential elements of the global climate which are most relevant to the operation of weather satellites.

2.1   Solar and terrestrial radiation

In its simplest terms the radiative balance of the Earth can be defined as follows: over time the amount of solar radiation absorbed by the atmosphere and the surface beneath it is equal to the amount of long-wave heat radiation emitted by the Earth to space. This basic balance does, however, cover a large number of physical processes which must be examined in order to understand both how the global climate functions and how weather satellites can observe the properties of the atmosphere and the surface beneath.

2.1.1   Radiation laws

It is a fundamental physical property of matter that any object not at a temperature of absolute zero ($-273\,°C$) transmits energy to its surroundings by radiation. This radiation is in the form of electromagnetic waves travelling at the speed of light and requiring no intervening medium. Electromagnetic radiation is characterised by its wavelength which can extend over a spectrum from very short gamma (γ) rays, through X-rays and ultraviolet, to the visible, and on to infrared, microwaves, and radiowaves. The wavelength of visible light is in the range 0.4–0.74 μm $(1\,\mu\text{m} = 10^{-6}\,\text{m})$.

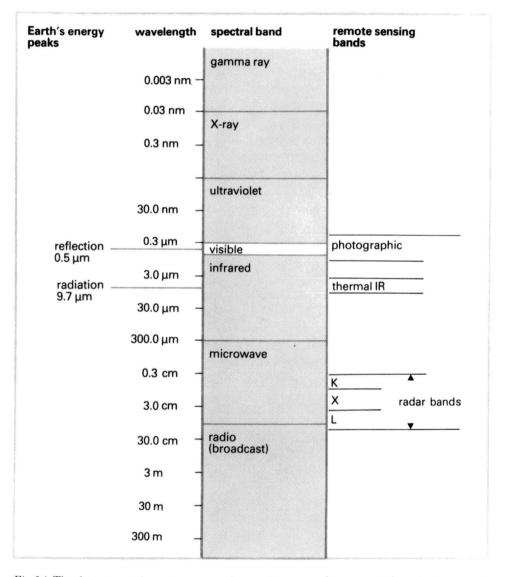

Fig. 2.1. *The electromagnetic spectrum covers the complete range of wavelengths from the longest radio waves to the shortest γ-rays.*

Throughout this book electromagnetic radiation will be identified in terms of its wavelength (generally in microns (μm) in the visible and infrared, and millimetres (mm) in the microwave region). There are, however, times when it is more convenient to discuss the frequency of the radiation. The relationship between the wavelength and the frequency is:

$$c = \lambda f$$

where λ is the wavelength in metres, f is the frequency in cycles per second, and c is the velocity of light (3×10^8 m s^{-1}).

The following units of frequency are most commonly used in satellite meteorology: in the microwave region, gigahertz (GHz) where 1 GHz $= 10^9$ Hz and 1 Hz is 1 cycle per second (cps); and in the infrared, cm^{-1}, which is a simple way of stating the number of wavelengths in 1 centimetre, where 1 cm$^{-1} = 3 \times 10^{10}$ Hz or 30 GHz. This means that when talking about microwave measurements scientists will use the wavelength or frequency interchangeably (eg., 5 mm or 60 GHz radiation), and similarly in the infrared (eg., 10 μm or 1000 cm^{-1}).

A body which absorbs all the radiation and, which at any temperature, emits the maximum possible amount of radiant energy is known as a 'black body'. In practice no actual substance is truly 'black'. Moreover, a substance does not have to be black in the visible light to behave like a black body at other wavelengths. For example, snow absorbs very little visible light but is a highly efficient emitter of infrared radiation. But at any given wavelength if a substance is a good absorber then it is a good emitter, while at wavelengths at which it absorbs weakly it also emits weakly.

The wavelength dependence of the absorptivity and emissivity of a gas, liquid, or solid is known as its spectrum. Each substance has its own unique spectrum which may have an elaborate wavelength dependence. This means that the radiative properties of the Earth are made up of the spectral characteristics of the constituents of the atmosphere, the oceans, and the land surface. These combine in a way that is essential to explaining the behaviour of the global climate.

For a black body, which emits the maximum amount of energy at all wavelengths, the intensity of radiation emitted and the wavelength distribution depend only on the absolute temperature. The expression for emitted radiation is defined by the Stefan–Boltzmann law which states that the flux of radiation from a black body is directly proportional to absolute temperature. That is:

$$F = \sigma T^4$$

where F is the flux of radiation, T is the absolute temperature as measured in K from absolute zero (-273.16 °C), and σ is a constant (5.670×10^{-7} W m^{-2} (K)^4).

It can also be shown that the wavelength at which a black body emits most strongly is inversely proportional to the absolute temperature. Known as the Wien displacement law, this is expressed as:

$$\lambda_m = a/T$$

where λ_m is the wavelength of maximum energy emission in metres and a is a constant $(2.898 \times 10^{-3} \, \text{m K})$.

2.1.2   Solar radiation

The Sun behaves rather like a black body with an effective surface temperature of 5750 K. This means that its maximum emission is in the region of 0.5 μm near the middle of the visible portion of the electromagnetic spectrum. Almost 99% of the Sun's radiation is contained in the short wavelength range 0.15–4.0 μm. Some 9% falls in the ultraviolet, 45% in the visible, and the remainder at longer wavelengths. In the atmosphere much of the ultraviolet and infrared radiation is absorbed. Ozone and oxygen are important absorbers at short wavelengths, while water vapour and carbon dioxide (CO_2) are the principal absorbers in the near infrared.

2.1.3   Terrestrial radiation

The average temperature of the Earth is 285 K, and so it emits radiation at longer wavelengths in the mid-infrared. Most of the energy is emitted in the range 4–50 μm with the peak near 15 μm. The shape of this curve exerts a strong influence on how the amount of energy emitted at given wavelengths varies as a function of temperature. At all wavelengths the amount of emitted energy rises as the temperature increases. But because the peak also shifts to shorter wavelengths as the temperature rises the amount of energy radiated at the shorter wavelengths rises more rapidly. This is important when trying to measure temperatures by observing changes in emitted radiation. A 1 K rise produces approximately a 1% increase in radiant energy in the 15 μm region often used for remote sensing. By comparison, in the microwave region (5 mm) the same temperature increase will produce only a 0.3% rise whereas in the near-infrared (4 μm) it produces about a 4% rise.

At any given point the amount of outgoing radiation from the Earth is proportional to both the temperature and emissivity of the surface, and the characteristics of the intervening atmosphere. Because the principal atmospheric gases (oxygen and nitrogen) do not absorb appreciable amounts of infrared radiation, the radiative properties of the atmosphere depend on the trace gases present (eg., water vapour, carbon dioxide, and ozone). Each of these interacts with infrared radiation in its own way. So the surface radiation is modified by absorption and re-emission in the atmosphere by these trace constituents whose temperatures will generally be different from that of the surface. Because this upwelling energy comes from both the surface and the atmosphere it is often referred to as 'terrestrial radiation'.

The effect of the radiatively active trace gases on the form of the terrestrial radiation is complex. Each species has a unique set of absorption and emission properties known as its 'molecular spectrum' which is the product of its molecular structure. This spectrum is made up of a large number of narrow features, often referred to as 'spectral lines', which tend to be grouped in broader bands around certain wavelengths. For different molecules the spectral lines and bands are at different wavelengths. The intensity of individual spectral lines and broader bands varies in a way that is related to the physical properties of each gas. So an analysis of the spectrum of either terrestrial radiation or solar radiation will pick out character-

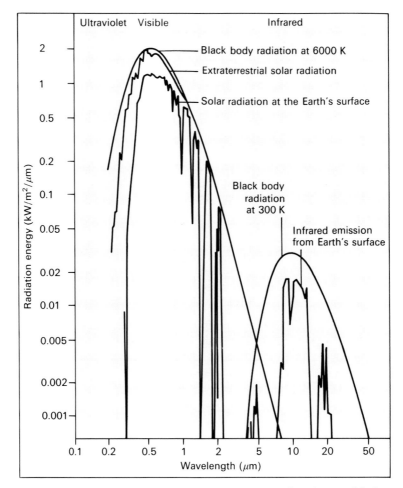

Fig. 2.2. The energy from the Sun is largely concentrated in the wavelength range 0.2–4 μm.

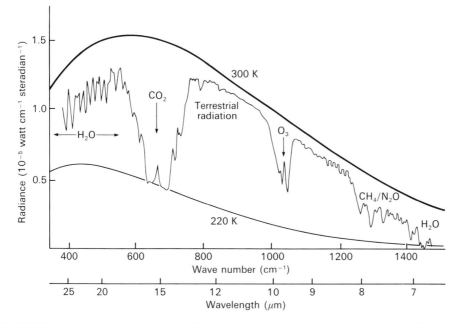

Fig. 2.3. The energy emitted by the Earth (terrestrial radiation) is concentrated in the 4–25 μm region and the amount emitted depends on both the temperature and trace constituents in the atmosphere.

istic features arising from the presence of the radiatively active gases. The way in which long-wave radiation either escapes to space or is re-absorbed by the atmosphere or the Earth's surface holds the key to temperature sounding from satellites (see Section 4.3.2).

2.1.4   The energy balance of the Earth

Before we can consider the detailed features of terrestrial radiation, it is necessary to look more closely at the way in which the balance with incoming solar radiation is achieved. At the local level, the total incoming flux of solar radiation is balanced by the outgoing flux of both solar and terrestrial radiation. This depends on various processes. Some solar radiation is either absorbed or scattered by the atmosphere and the particles and clouds in it. The remainder is either absorbed or reflected by the Earth's surface. The amount of energy absorbed or reflected is dependent on the surface properties. Snow reflects a high proportion of incident sunlight, while moist dark soil is an efficient absorber.

The amount of solar radiation reflected or scattered into space without any change in wavelength is defined as the albedo of the surface. The mean global albedo is about 30%. The albedo of different surfaces can vary from 90% to less than 5%. Examples of the albedo of different surfaces are given in Table 2.1. Since roughly

Table 2.1. *Albedo of various surfaces*

Nature of surface	Albedo
Stable dry snow cover (latitudes above 60°)	0.80–0.90
Forest with stable snow cover	0.45
Sea ice (slightly porous milky-bluish)	0.38
Sand, bright fine	0.37
Deciduous forest in summer	0.18
Ploughed field, moist	0.14
Oceans (latitude 60°)	0.07–0.20

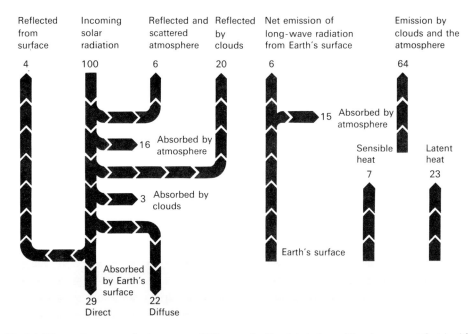

Fig. 2.4. *The total amount of solar energy falling on the Earth is balanced by the amount that is either reflected or absorbed and re-emitted as terrestrial radiation.*

half the Earth's surface is obscured by clouds, their contribution to the overall albedo figure is also important. The albedo of different types and levels of cloud is roughly proportional to cloud thickness. Values for various types of clouds are given in Table 2.2.

As about 30% of incoming solar radiation is reflected or scattered back into space, the remainder must be absorbed. Of the total incoming flux about half penetrates the atmosphere and is absorbed by the Earth's surface. The rest (some 16%) is absorbed directly by the atmosphere. Both the atmosphere and the surface re-radiate

Table 2.2. *Albedo of clouds*

Type of cloud	Albedo
High-level clouds (cirrus)	0.21
Middle-level cloud sheets (3–6 km)	0.48
Low-level cloud sheets	0.69
Cumulus clouds	0.70

this absorbed energy as long-wave radiation. But this is not the whole story. Because the atmosphere and oceans transport energy from one place to another this motion is an integral part of the Earth's radiative balance. On a global scale the most important feature is that at high latitudes, despite lower temperatures (typically 240 K in polar regions in winter), large amounts of energy is radiated to space. This process is greatest in winter, when there is little or no solar radiation reaching these regions. This loss must be balanced by energy transport from lower latitudes. Even though this energy transfer is reduced at other times of the year, it provides the energy to fuel the engine which drives the global atmospheric and oceanic circulation throughout the year.

2.2   The global atmosphere

The global atmosphere is immensely complicated. It has structured behaviour at every scale and Swift's observation about fleas has been parodied by the British meteorologist Lewis Richardson to explain this complexity: 'Big whirls have little whirls that feed on their velocity / And little whirls have lesser whirls, and so on to viscosity'. This endless succession of vortices is evident in many satellite images. But to understand the information that can be obtained by viewing the atmosphere from space, we must review the entire range of motions that make up the weather. Before doing so it is essential, however, to bear in mind at all times the fact that these motions have a vertical dimension. It is easy to lose sight of this fact when looking at two-dimensional images. But without an awareness of the simultaneous vertical movement of the atmosphere, much of the understanding of weather systems is lost.

2.2.1   *General circulation*

The weather features we have to consider here range from local breezes and shower clouds to the great wave patterns that circle the globe. All of these systems are part of the process of the atmosphere transport of energy. At the largest scale this is a reflection of the fact that the solar energy absorbed in equatorial regions is greater than the outgoing infrared radiation, whereas in polar regions the reverse

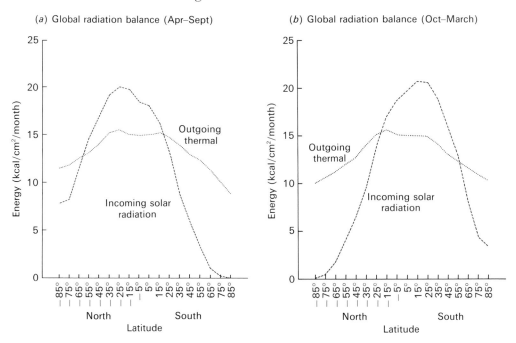

Fig. 2.5. In general the amount of solar energy absorbed by the Earth at each latitude differs from the amount of terrestrial radiation emitted at the same latitude and so energy has to be transferred from equatorial to polar regions.

applies. So, to achieve a global energy balance the atmosphere and oceans must transport energy from the equator to the poles. This is the engine that drives the principal components of the global weather machine.

If the Earth's surface were smooth and of uniform composition, the long-term mean patterns of wind, temperature, and rainfall would show nothing but zonal bands with no longitudinal variation. Moreover, if the Earth did not rotate energy transfer would involve a simple meridional circulation with air rising at the equator and flowing to the poles, descending, and returning at low level to the equator. With a rotating Earth the motion involves horizontal vortices and waves. The distribution of the oceans and continents across the globe make these motions still more complex, but the broad features retain many of the features of the simple zonal models.

A schematic representation of the zonal wind systems near the Earth's surface consists of easterly trade winds in the tropics, calm subtropical high pressure zones, mid-latitude westerlies, and stormy low and high pressure zones close to the poles. In cross-section these zones can be represented as three circulation systems covering the tropics, the mid-latitudes, and the polar regions. These general circulation features, both horizontal and vertical, are clearly recognisable from satellite images.

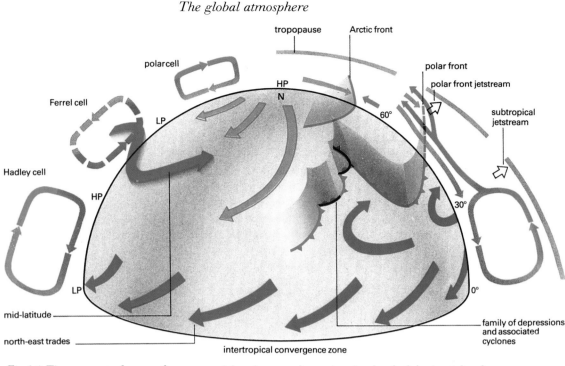

Fig. 2.6. The transport of energy from equatorial regions to polar regions involves both horizontal and vertical circulation of the atmosphere.

The weather in tropical regions is dominated by vertical circulation. Known as the Hadley cell, after George Hadley, the eighteenth-century London lawyer who first proposed it, the general motion consists of air rising near the equator to heights of up to 20 km and then spreading out to 20–30° N and S latitudes before descending and flowing back toward the equator. The rising air at the equator is humid, and cools as it rises. This leads to the formation of towering shower clouds which girdle the Earth and produce heavy rainfall in equatorial regions. The precise position of this band of convective activity, known as the 'Intertropical Convergence Zone' (ITCZ), varies with the movement of the noonday Sun throughout the year. It tends to follow the Sun, bringing summer rains to the drier regions around 20° N and 20° S. Its movement is also linked to the distribution of the oceans and continents in the equatorial regions. Understanding this motion is vitally important, as the movement of the ITCZ brings life-giving rain to over two billion people living in the tropics.

The descending air in the Hadley cell is dry and warm and hence the regions around 20–30° N and 20–30° S latitudes are arid. These desert regions stand out clearly on satellite images and confirm how well the simple model proposed by Hadley explains the broad features of the weather in the tropics. But this basic picture disguises subtle variations of huge significance. The long-standing drought in the Sahel region of sub-Saharan Africa which is a result of the failure of the ITCZ to move as far northward as normal since the late 1960s, is the best example of such shifts. More generally, fluctuations in the intensity of this tropical weather regime are fundamental factors in global climate changes. So long-term satellite monitoring of the behaviour of the Hadley cell is integral to understanding such changes.

The circulation of the tropics can be described in relatively simple terms because variations in surface temperature are relatively small. But in extratropical regions global circulation of the atmosphere is controlled by a different set of conditions. These are characterised by strong rotational motion which involves substantial temperature contrasts. At low altitudes this motion involves mobile areas of low pressure depressions and high pressure anticyclones. Above about 3 km altitude it consists primarily of large waves, generally moving from west to east. These are superimposed upon a strong zonal current, the core of which is called the 'jet stream'.

The existence of long waves in the upper atmosphere has a fundamental influence on the weather patterns of the mid-latitudes, as they tend to guide the course of the low-level eddies. The fact that these waves are the consequence of the combination of the Earth's rotation and temperature contrast can be demonstrated in the laboratory. A rotating dishpan heated on its outer wall and cooled at the centre can be made to produce either symmetric or standing wave patterns depending on the rate of rotation and the temperature difference across the fluid. These waves can be shown to be responsible for the main heat transfer between the warm rim and the cool centre.

In the real world the patterns are much more complex, reflecting both the changes of the seasons and the distribution of the oceans and continents. In particular, these factors exercise great influence in the Northern Hemisphere. In winter the jet stream is strongest near the east coast of Asia and over the eastern United States and North Africa where strong temperature gradients prevail. In the summer the jet stream is much weaker and is located further north.

At sea level the mean pressure pattern reflects even more dramatically the influence of the oceans and continents. In addition the effect of the major mountain ranges is crucial. In winter, low pressure areas form over the warm oceans and high pressure areas over the cold continents. So, the pattern is dominated by low pressure areas over the North Pacific and Iceland, and a high pressure area over Siberia. The high over Siberia is displaced far to the east as there is no barrier to Atlantic air moving in from the west, whilst the mountains to the south prevent any appreciable exchange with the Pacific and southern continents. Over North America the high nestles in the lee of the Rocky Mountains. In summer the pattern shows less contrast and is effectively reversed with the high pressure over the relatively cool oceans and lower pressure over the warm continents.

While weather patterns in the mid-latitude zone (35–55° N) can be characterised by these seasonal averages, at any one time the patterns can show a huge variety. Within this constant flow of differing daily weather types it is possible to show that these patterns exhibit irregular quasicyclic changes which can best be measured in terms of pressure difference across the zone. These shifts are between regimes of high and low zonal pressure difference. With a large pressure difference there are strong sea-level westerlies and a long wavelength pattern in the upper atmosphere. With a low pressure difference there is a breakdown of the sea-level westerlies into closed cellular patterns and a corresponding meandering shorter wavelength pattern aloft. The breakdown of the westerlies is particularly interesting as it can lead to prolonged periods of abnormal weather with compensating extremes around the globe.

Before moving on to specific weather systems there is one additional feature of general circulation of the Northern Hemisphere which deserves special mention. This is the monsoon over the Indian subcontinent. The basic explanation was first proposed by Edmond Halley to the Royal Society in 1686. He suggested that as Asia warms up in the summer it has the effect of drawing huge quantities of moisture northward from the tropical Indian Ocean. In the winter the reverse occurs with the winds blowing from the cold continent to the warm sea. This is, however, only part of the story. The reason that the summer monsoon is so much stronger over India than elsewhere in the tropics is the controlling influence of the Himalayas and the Tibetan plateau. As Asia heats up, the jet stream switches from south of the Himalayas to northern Tibet, and the moist tropical air and heavy rains associated with the ITCZ are drawn farther north over India than in other parts of the world.

In the Southern Hemisphere circulation is easier to explain for two reasons: first, in the mid-latitudes the near absence of land masses greatly reduces the contrasting effects of the oceans and continents; secondly, the almost symmetrical distribution of Antarctica about the South Pole reinforces the zonal circulation of the atmosphere. So at both sea level and aloft the flow is more uniform. The westerlies are stronger and show less variation between winter and summer. Also pressure differences across the mid-latitude zone are less marked. This means there is less tendency for the zonal flow to break down into a meandering pattern that produces abnormal weather.

2.2.2 Fronts, depressions, and anticyclones

These general observations about circulation provide only a basic impression of the climatological behaviour of the global weather machine. This is an average of a continual series of dynamic systems which produce the variety we know as 'the weather'. In mid-latitudes the essential feature of this variability is the succession of fronts, depressions, and anticyclones. It is satellite images of these weather

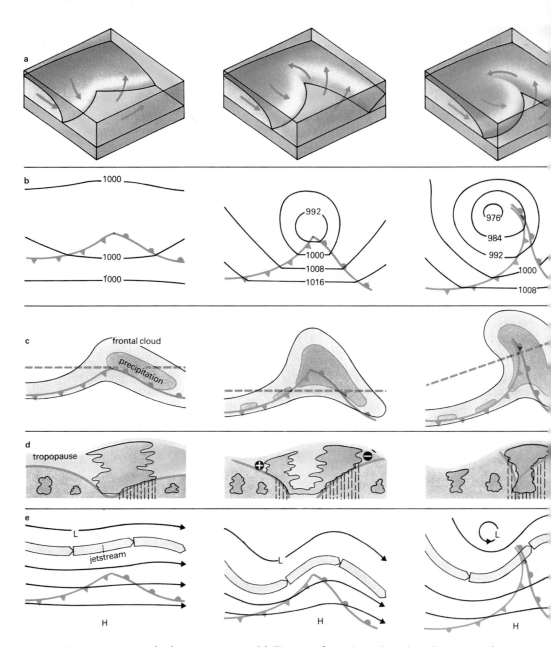

Fig. 2.7. The development of a frontal depression: (a) The view from above shows how the wave on the frontal surface grows and finally occcludes over a period of days. (b) shows the warm and cold fronts on the ground and associated surface atmospheric pressure distribution. (c) shows the distribution of frontal clouds and precipitation. (d) indicates the cloud development and precipitation in relation to the vertical structure of the frontal zones along the cross-sections indicated in (c). (e) shows the position of the fronts at the ground and the flow aloft including a jet stream axis which is typically associated with a frontal depression. The middle diagram of (d) shows the jet stream entering the page (+) and leaving the page (−).

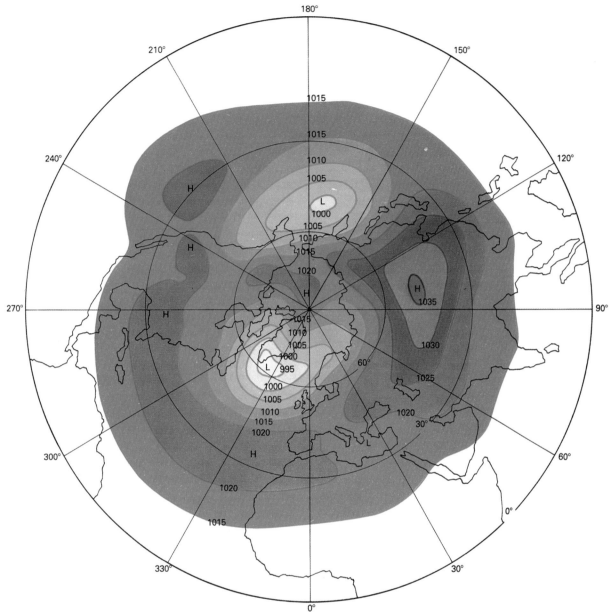

Fig. 2.8 Average sea-level pressure in millibars in the northern hemisphere in January. The subtropical high-pressure cells are permanent features, and the low-pressure areas in mid-latitudes are the result of the frequent passage of travelling depressions.

features that are most often published in papers, magazines, and on television, and so we need to know something about their physical properties.

In 1863, the first head of the Meteorological Department of the British Board of Trade (the forerunner of the Meteorological Office), Admiral Fitz-Roy, proposed a model of the essential feature of an extratropical depression. He observed that depressions were normally composed of two air masses of different temperatures and moisture content, with the warmer, more moist of these originating in subtropical latitudes while the colder drier mass came from polar regions. In 1919 the Norwegian meteorologist Jacob Bjerknes published a model of depressions that is still largely used today. In this and subsequent work, he defined the structure and dynamic processes of a simple mobile depression, and described how it went through a series of stages and propagated, often as one of a family, along the front between polar and subtropical air.

The important feature of the model is that the separation between the two air masses is normally sharp. Along this front temperature and wind change suddenly, and in most cases the state of the sky changes likewise. Sections of the front are named according to their motion; a warm front is a front along which warm air displaces cold air; a cold front has the reverse characteristic. A cross-section of a typical depression to the south of its centre shows wedges of cold air under warm air. To the north of its centre the cold air forms a trough filled with warm air. The overall effect of the depression is to raise the warm air creating extensive layers of cloud and precipitation in the region of the warm and cold fronts.

The life cycle of a depression, as proposed by Bjerknes, starts with a quasistationary front between a cold and warm air mass with a shearing motion between them. This develops into a wave which eventually winds up as a well-developed depression. In this process the cold front catches up with the warm front using up the wedge of warm air between them to drive the depression through its development. Where the cold front overtakes the warm front the system is said to 'occlude' and the resulting 'occluded front' has no temperature change at ground level but may be marked by heavy precipitation. In its later stages the depression runs out of energy and begins to stagnate, but its trailing cold front can become the site for the next 'secondary' depression. In this way a family of several depressions can follow one another along similar tracks.

This model successfully describes the properties of extratropical depressions, but it cannot explain the behaviour of fronts. Their remarkable feature is that air flow should generate such sharp zones with horizontal scales of only a few tens of kilometres, rather than maintaining a weaker, broader thermal contrast between the equator and pole. Moreover, the model does not handle the basic question of which comes first, the front or the depression. In many cases the depression seems to come first, contrary to the Bjerknes model. These problems are now the subject of intense investigation combining aircraft, radar and surface measurements, satellite cloud observations, and computer models. This work is aimed at identifying the nonlinear feedback mechanisms which can produce the observed near-discontinuity at the front. In time it may lead to improved forecasting of the rain and snowfall that are concentrated at fronts.

By comparison with the dynamic and apparently well organised behaviour of depressions, anticyclones are more enigmatic features. They lack the same degree of uniformity in their patterns of formation, growth and decay, being far more irregular in behaviour as well as in shape. Often they appear as sluggish, passive systems which fill the spaces between their vigorous cyclonic counterparts. They can be divided very broadly into 'cold' or 'polar' continental highs and 'warm' or 'dynamic' highs.

Polar continental highs develop over the northern land masses in winter. They are created by intense cooling of the snow-covered surface which gives rise to a shallow layer of dense very cold air. Dynamic anticyclones are caused by large-scale subsidence throughout the depth of the lowest 10 km or so of the atmosphere which is usually known as the 'troposphere' (see Section 4.3.1). These include the subtropical highs associated with the subsidence of the descending limb of the Hadley cell, and slow-moving mid-latitude highs that are known as 'blocking anticyclones'. The latter are closely associated with the movement of depressions and the pattern of long waves in the mid-latitudes.

At the simplest level dynamic highs are the areas of high pressure that punctuate a family of depressions. They may be no more than ridges of high pressure which glide along the border of a much larger subtropical anticyclone. But often they can assume a separate identity and play a much more influential role in affecting weather in the mid-latitudes. In particular, when they become stationary and form an integral part of the establishment of a global pattern of long waves which persists for several weeks, then they become a dominant force. These 'blocking anticyclones' play a central part in spells of abnormal weather around the world.

As noted earlier, the strength of the mid-latitude westerlies fluctuates between a strong long wavelength pattern and a meandering short wavelength pattern. The latter is frequently associated with the establishment of blocking conditions. These are defined by the upper-level westerlies being split into two well-defined branches which extend over at least 45° of longitude. To constitute a block, these cells of high pressure, which are sandwiched between normal zonal westerly flow, must last at least 10 days. This condition is frequently met, as once blocks form they are relatively stable persisting on average for 12–16 days, although they can last much longer. In the Northern Hemisphere their position is influenced by the distribution of the continents and mountain ranges. They most frequently occur close to the Greenwich meridian and in the eastern Pacific. Atlantic blocks are approximately twice as common as the Pacific variety. In the Southern Hemisphere blocking is less common – a fact that may be related to the relative absence of land masses in the southern mid-latitudes.

2.2.3   Tropical storms and hurricanes

While general circulation in the tropics is relatively simple because of the small regional variations in the surface temperature, they can spawn storms which sweep outward to higher latitudes. Known variously as cyclones, hurricanes, and typhoons, these storms play an important part in the transport of energy away from the tropics, while doing a huge amount of damage on the way.

Tropical storms have their genesis in weak wave features and minor depressions that occur along the edge of the ITCZ. Many of these features live out their lives as minor disturbances of weakly organised convective activity, but some of them develop into major storms. What causes one disturbance to come to nothing and another to grow explosively is not fully understood. What is well known is that once the conditions are right the storm grows rapidly and frequently follows a standard pattern. As the pressure starts to fall rapidly at the centre of the disturbance, the winds rise in a tight band of some 30–60 km radius and the clouds form into a remarkably circular pattern about a central eye. The circular cloud system then

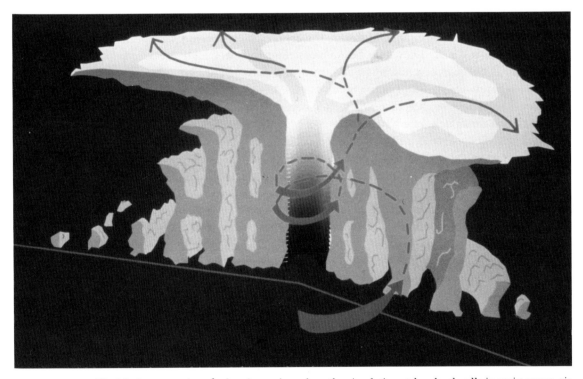

Fig. 2.9 A cross-section of a hurricane shows how the circulation at low level pulls in moist warm air which rises up in the centre of the storm and spreads out at high levels (with permission of NASA).

22

expands until the storm reaches full maturity. The maximum wind speed around a storm is highly correlated to the degree of organisation and the diameter of the circular area.

As the storm grows, it moves westward in the trade winds (usually in the belt 8–15° from the equator) and then starts to migrate toward higher latitudes. The maturing storm then expands while the central pressure stops falling. It tends to follow an unpredictable path which depends on the surface conditions. In particular sea surface temperature is a vital ingredient. Hurricanes only develop over water

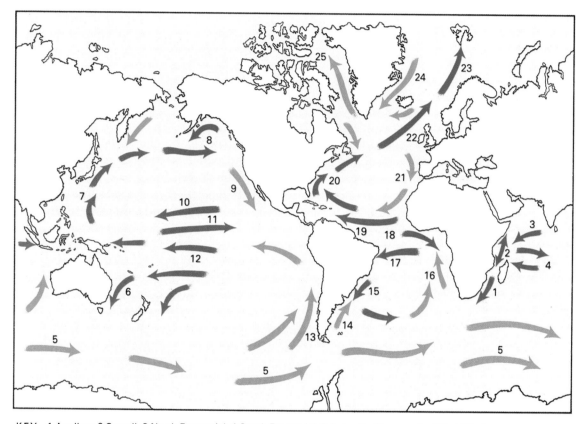

KEY 1 Agulhas, 2 Somali, 3 North Equatorial, 4 South Equatorial, 5 Antarctic Circumpolar (West Wind Drift), 6 West Australian, 7 Kuroshio, 8 North Pacific/Alaska, 9 California, 10 North Equatorial, 11 Equatorial Counter, 12 South Equatorial, 13 Peru, 14 Falkland, 15 Brazil, 16 Benguela, 17 South Equatorial, 18 Equatorial Counter, 19 North Equatorial, 20 Gulf Stream, 21 Canaries, 22 North Atlantic Drift, 23 Norwegian, 24 East Greenland, 25 West Greenland.

Fig. 2.10. The ocean currents also play a major role in the transport of energy from equatorial to polar regions. The figure shows the major warm and cold currents. © *Sceptre Books, a division of Time Life International Limited, London*

with a temperature above 27 °C and their subsequent course is closely related to where the warmest water is. The storm can grow to a radius of more than 300 km before it starts to decay. The declining phase is accelerated by passing over either land or colder water. By this time the hurricane is not only moving toward higher latitudes but is also being pushed eastward by the mid-latitude westerlies. The winds, by now, have lost much of their intensity, but the dying storm can still produce towering clouds, heavy rainfall, and flooding.

To qualify as a hurricane, a tropical storm must have sustained winds around the eye in excess of 120 km h^{-1}. In extreme cases hurricanes have had wind speeds of over 300 km h^{-1}. Satellite imagery has confirmed that the degree of organisation and symmetry of a storm is linked to its intensity – the higher the degree of concentricity the more vigorous the storm. The damage caused by storms is principally a consequence of the high winds. In addition, the winds pile up water ahead of the storm and when it makes landfall this 'storm surge' can do immense damage to low-lying coastal areas. In the later stages of the storm heavy rainfall can become the most damaging element.

Most attention is devoted to the detailed aspects of the behaviour of individual tropical storms and hurricanes. But we must not lose sight of the fact that they are an integral part of the global energy transport from the equator to the poles. They occur in the tropical oceans of both hemispheres, being most common in the North Pacific. They are most frequent in late summer when the sea surface temperatures are highest, but can occur throughout the summer and autumn in the tropics.

2.3 Ocean currents

So far we have concentrated on the role of the atmosphere in transferring energy poleward. Equally important, though less well understood, is the part played by the oceans in this process. The broad features of the circulation of the atmosphere and the oceans are closely linked. The basic form of ocean circulation in both hemispheres is an easterly flow close to the equator followed by a move toward higher latitudes along the western margin of the ocean basins. At latitudes above about 35° N and S the major currents sweep eastward carrying warm water to higher latitudes. This pattern is seen most clearly in the North Atlantic and North Pacific in the form of the Gulf Stream/North Atlantic Current and the Kuroshio/North Pacific Current. To balance this poleward flow there are returning currents of cold water moving down the eastern side of the ocean basins. The same type of compensating motion operates in the Southern Hemisphere but because of the virtual absence of land between 35 and 60° S the flow is combined with a strong circumpolar current round Antarctica.

The broad features of these currents have been known to mariners for centuries and to geographers for over 100 years. But detailed knowledge of them remains patchy. Their structure, the amount of energy they transport, and their changes from year to year are still largely a mystery. This is because oceanographic studies using ships and buoys can only provide a fragmentary picture of the underlying processes. Satellite observations have already provided some important insights into the large-scale behaviour of ocean currents.

2.4   Temperature regimes

The overall result of global atmospheric and oceanic motions is to establish the climatic zones that govern our lives. Many of the details of these will be explored in later chapters, but we need to summarise the way in which the temperature regimes associated with the various climatic zones reflect broad movements of the atmosphere and oceans. These effects show up most clearly in the Northern Hemisphere in winter. The temperature distribution in January is dominated by the meridional variation in net balance between incoming solar radiation and outgoing longwave radiation. On this pattern the effects of the atmosphere, oceans and continents are superimposed. In the tropics the temperature is uniform and falls only slowly northward so that the 20 °C isotherm runs around the globe close to 20° N latitude. But increasing deviations occur at higher latitudes with the 0 °C isotherm dipping to 40° N over North America and 35° N over East Asia while rising to 70° N off the coast of Norway and 60° N in the Bay of Alaska.

These variations result principally from the combination of the high heat capacity and motion of the oceans compared with the low heat capacity and efficient radiative properties of the snow-covered areas of Asia and North America. So the lowest winter temperatures occur over the heart of the continents. However, prevailing winds and travelling weather systems also affect temperatures. In western Europe the westerly winds from the Atlantic carry the 0 °C isotherm far inland. On the other side of the Atlantic it lies well out from the east coast because the weather systems draw cold continental air out of North America in their wake.

In summer temperature differences between the equator and the pole are much reduced, as are contrasts between the oceans and the continents. Longitudinal differences in July are caused by the heating of land masses by solar radiation. The most significant features are high temperatures over the deserts of North America, North Africa, the Middle East, and Central Asia. The oceans heat up much less quickly and so are in general cooler than adjacent land masses.

In the Southern Hemisphere the variations are less striking. In low- to mid-latitudes cold currents reduce the temperatures along the western coasts of South America and southern Africa, but the effects are small compared to the Northern Hemisphere. The absence of continents at high latitudes and the circumpolar form of Antarctica means that the temperature regime at these levels is almost uniform

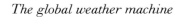

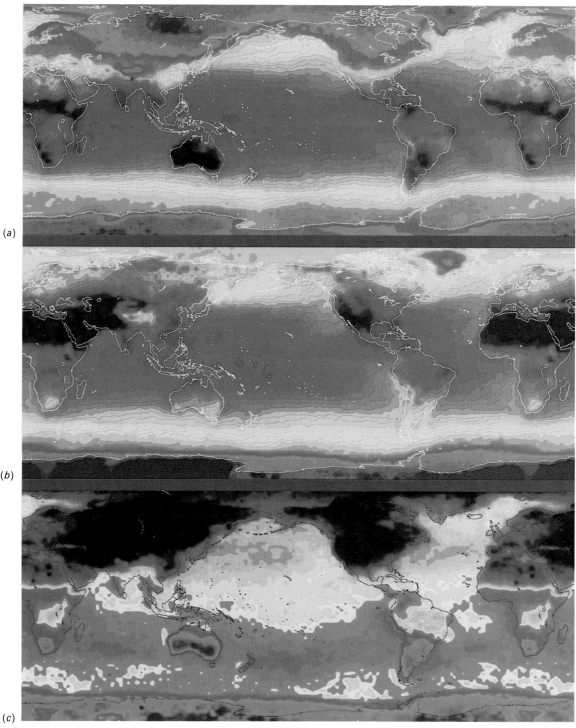

Fig. 2.11. A compilation of satellite observations showing typical surface temperatures of the Earth during (a) *January,* (b) *July and* (c) *the difference between July and January (with permission of NASA).*

with longitude. The annual cycle is dominated by radiation changes and the resultant expansion and contraction of the pack ice around Antarctica. This huge annual pulse has the effect of creating significant changes in temperatures at high latitudes with no appreciable longitudinal variation.

All of these broad features of global surface temperatures have been recorded by satellites and are encapsulated in colour images. Not only have these images been used to provide a graphic illustration of the seasonal patterns described above, but they can also provide a measure of the annual cycle of temperatures. For example, satellite images have shown that there are seasonal temperature changes of up to 30 °C in both hemispheres with the greatest differences being in the interiors of the northern continents.

3

The development of satellite meteorology

*'Nothing in progression can rest on its original plan.
We may as well think of rocking a grown man
in the cradle of an infant.'*
Edmund Burke, 1729–97

WEATHER SATELLITES are but the latest and most obvious step in the development of meteorological observing systems. From the earliest days meteorologists have been developing equipment to address practical problems. The first weather service in Britain was set up in part as a response to the tragic loss of life when the passenger liner *Royal Charter* sank in 1859 in an unexpected storm off the coast of north Wales. In North America the needs of agriculture were a major stimulus to providing forecasting and climatological information. Later the demands of aviation, especially during World War II and after, were the driving force for greater understanding of the behaviour of the upper atmosphere. Satellite meteorology is the logical extension of these earlier pressures to develop better measurement systems.[1]

The prospect of viewing the weather from space has long fascinated scientists. In 1860, in his book *From the Earth to the Moon*, Jules Verne described how the 'lunarnauts' looked back to see 'cloud systems against the Earth's background' with 'some parts brilliantly lighted showed the presence of high mountains, often disappearing behind thick spots which were never seen on the lunar disc. They were rings of clouds placed concentrically round the terrestrial globe.' But it was not until the development of rocket technology in World War II that the prospect of viewing the weather from above became a reality.

3.1   The early days

The first successful photographs of cloud systems from space were obtained in 1947 using a converted V2 rocket at an altitude of between 110 and 165 km. This led to the first of a series of proposals about obtaining useful pictures from space, and work began on concepts for satellites. By the early 1950s it was recognised that rockets could produce useful information. In particular, meteorologists had not anticipated the extraordinary complexity of weather systems which the photographs revealed.

In July 1955 Professor Joe Kaplan, Chairman of the US International Geophysical Year Committee, announced in Brussels that President Eisenhower had agreed to launch a satellite as part of the US contribution to the year. Spurred on by the launch of Sputnik in October 1957 the President announced the creation of the National Aeronautics and Space Administration (NASA) in March 1958. The first 'meteorological' satellite was the First Earth Radiation Experiment which was launched on 13 October 1958 and made useful measurements of the radiation balance of the Earth.

3.2 Global atmospheric research

The era of satellite meteorology truly began with the launch of the first operational satellite – Television Infra Red Observational Satellite (TIROS I) – on 1 April 1960. While the success of TIROS I demonstrated the new insights to be gained from viewing the weather from space, it produced more questions than answers. During its operational life of 89 days it transmitted 23 000 pictures, over half of which were meteorologically useful. These first pictures showed both the limitations of the system, and simultaneously, gave scientists dramatic proof of the complexity of the weather processes that needed to be measured. As had been recognised from the outset, an operational network of weather satellites to meet the needs of meteorologists required a whole set of technological developments that involved advances over a broad scientific front.

At the same time it was realised that not only would the development of meteorology have to be built around satellite systems but also digital computers would play a central role in both data handling and, more importantly, the 'number crunching' required for better forecasts. In early 1961 the US National Academy of Sciences issued a report describing the prospects for the development of atmospheric sciences through the use of meteorological observations from space. On the basis of those findings and at the initiative of the United Nations Committee on the Peaceful Uses of Outer Space, the United Nations General Assembly adopted a resolution in December 1962 on international cooperation for the development of meteorology and atmospheric sciences which placed particular emphasis on the use of meteorological satellites. The two major planks of the subsequent international effort – World Weather Watch and the Global Atmospheric Research Programme (GARP) – were firmly based on measurements from space, and defined a set of

Table 3.1. *Observational requirements for the GARP*

Atmospheric state parameters	Accuracy route mean square
Wind components	$\pm 3\,\mathrm{m\,s^{-1}}$
Temperature	$\pm 1\,°\mathrm{C}$
Pressure of reference level	$\pm 0.3\%$
Water vapour pressure	$\pm 1\,\mathrm{mbar}$
Sea surface temperature	$\pm 0.25\,°\mathrm{C}$
Space and time averages	
Time average interval	2 h
Horizontal space average	100 km
Vertical space average	Defined by the requirement of a minimum of 8 data levels at surface – 900, 700, 500, 200, 100, 50, 10 mbar, respectively
Further information for verification purposes	
Precipitation	
Cloud cover data	
Snow or ice cover	
Elements of radiation budget	

requirements which remain the yardstick by which to gauge the success of modern satellite systems (Table 3.1).

This far-seeing approach was not, however, immediately accepted by many practising weather forecasters who considered that the data and images from TIROS I and subsequent satellites in this series did not fit in with their standard ground-based measurements. This inertia had to be overcome by a series of developments and a growing acceptance of the value of the data.

3.3 Satellite programmes

The dominant force in the development of satellite meteorology has been the United States. Throughout this book the National Aeronautics and Space Administration (NASA) occupies a central place. However, other major programmes have run in parallel with the American work. In particular, the Russians have built up a substantial network of orbiting satellites to provide complete coverage of the Soviet Union. Their first meteorological satellite, COSMOS 122, was launched in 1966. By the end of 1983 they had put over 40 weather satellites into orbit which they describe as being able to watch 'high speed natural phenomena, paths of cyclones, tidal action, dust storms, and tsunamis'. They also claim that advance warning of tropical storms has saved many lives. Irrigation planning has been improved using measurements of snow cover to calculate the seasonal run-off from the Tien Shan Mountains and the Himalayas each summer. The sailing time of Soviet ships has been cut by 10% as a result of being able to avoid storms, adverse winds and heavy ice conditions.

As a general observation the Soviet systems are several years behind the American work. The first operational meteorological satellite of the Soviet Meteor 1 series was launched in 1969 and took only television pictures. The second series (Meteor 2) using scanning radiometers (see Section 4.2) came into service in 1975, whereas the equivalent US system had first flown in 1969. So, in general, the Soviet work will not be covered here. Other major contributors to the development of satellite meteorology include: the European Space Agency, whose geostationary satellites Meteosat 1 to 4 have made a major contribution to weather forecasting in the last decade; and countries such as India and Japan which have either launched their own satellites, or flown instruments in US satellites. All of these systems have contributed to the overall development of satellite meteorology.

The scale of these programmes was massive. They encompassed the development of improved communications systems and data handling methods, and huge advances in computer technology. At the same time, they posed a series of challenges to meteorologists who already had a well-established ground-based system for observing the global atmosphere. Grafting an entirely new set of data onto this existing body of knowledge required a great deal of effort. So, despite the obvious advantages of observations from space, the application of much of the data required a lengthy process of development.

In the case of pictures of weather systems, such as thunderstorms, fronts, depressions, and hurricanes, satellite meteorology was of immediate use to meteorologists. It provided confirmation of their analysis and could be used to reinforce broad features of forecasts. The most dramatic example of this was in the tracking of hurricanes whose erratic behaviour had for decades confused forecasters. By the late 1960s the US National Weather Service was using satellite images to predict the landfall of hurricanes. When Hurricane Camille, the most destructive hurricane on record for the US mainland, hit the Louisiana–Mississippi coast in August 1969, the loss of life was relatively small as most of the populace had heeded the warnings and taken shelter inland. Satellite images had played an invaluable part in this successful forecast.

While this qualitative use of satellite images could be of immediate aid to forecasters, the development of accurate measurements of temperature, humidity, pressure, wind speed, and rainfall took longer. In some cases it was possible to show that certain information, such as the amount of rainfall or the possible outbreak of tornadoes, could be inferred from the basic visible images. But for improved forecasting instruments were needed which could make reliable observations of meteorologically important parameters at regular intervals over the whole globe.

Fig. 3.1. Hurricane in the Bay of Bengal (with permission of NASA).

3.3.1 Polar-orbiting satellites

In providing the observation system needed, meteorologists had to draw on a variety
of satellite programmes. This effort was complicated by the two principal objectives
of the satellite programmes. The first was to put an operational observing system
into place as soon as practical. In the United States the major planks of this effort
were the TIROS series and its successors (ESSA, ITOS, and TIROS-N/NOAA series)
and the geostationary GOES series (see Section 3.3.2). The second was to develop
new measurement techniques and prove that they were capable of providing reliable
observations (see Section 3.3.3). Foremost among these programmes has been the
NASA Nimbus series. But other systems, notably the short-lived SEASAT in 1978,
have made major contributions to meteorology. The various programmes are sum-
marised in Tables 3.2 and 3.3.

There has been considerable overlap between the operational and experimental
systems. Operational satellites incorporated new instruments which were still at an

Table 3.2. *Satellites in TIROS/ESSA/ITOS/NOAA series*

Name	Launch date	Operational lifetime (days)	Comments
TIROS I	1 Apr 1960	89	Limited global coverage, TV
TIROS II	23 Nov 1960	376	pictures stored on board and
TIROS III	12 July 1961	230	transmitted to specific receiving
TIROS IV	8 Feb 1962	161	stations
TIROS V	19 June 1962	321	
TIROS VI	18 Sept 1962	389	
TIROS VII	19 June 1963	1809	
TIROS VIII	21 Dec 1963	1287	First direct read-out system
TIROS IX	22 Jan 1965	1238	First satellite providing global coverage
TIROS X	2 July 1965	730	
ESSA 1	3 Feb 1966	861	First operational system
ESSA 2	28 Feb 1966	1692	
ESSA 3	2 Oct 1966	738	Carried a variety of TV systems and
ESSA 4	26 Jan 1967	465	advanced image-forming devices
ESSA 5	20 Apr 1967	1034	which measured reflected sunlight
ESSA 6	10 Nov 1967	763	
ESSA 7	16 Aug 1968	571	
ESSA 8	15 Dec 1968	1103	
ESSA 9	26 Feb 1969	1030	
ITOS 1	23 Jan 1970	510	Equipped with automatic picture transmission and advanced image-
NOAA 1	10 Dec 1970	252	forming apparatus
NOAA 2	15 Oct 1972	837	Equipment included a Very High
NOAA 3	6 Nov 1973	1029	Resolution Radiometer (VHRR) and
NOAA 4	15 Nov 1974	1463	a vertical temperature profile
NOAA 5	29 July 1976	1067	radiometer
TIROS-N	13 Oct 1978	868	Equipment included Advanced Very High Resolution Radiometer
NOAA 6	27 June 1979	2560	(AVHRR), infrared and microwave
NOAA 7	23 June 1981	1235	temperature-sounding systems, and
NOAA 8	28 Mar 1983	533	a stratospheric sounding unit
NOAA 9	12 Dec 1984	*a*	
NOAA 10	20 Sept 1986	*a*	
NOAA 11	24 Sept 1988	*a*	

a In operation in early 1990.

early stage of development. Conversely, experimental equipment which provided successful results was often exploited by weather services. This makes it difficult to follow how each system contributed to the advance of satellite meteorology. The plethora of acronyms developed in the subject area, some of which are altered when control of the satellite is transferred from the launching authority (NASA) to the operating authority (eg., NOAA), further confuses the issue (see Acronyms on p.185). In this chapter we will concentrate on important technical landmarks over the last 25 years.

Dramatic as the early TIROS photographs were, the spacecraft configuration, orbit, and orientation imposed serious limitations. On the first 8 TIROS satellites the cameras pointed outward from the rim of the spinning spacecraft so that they only viewed the Earth for about 25% of the time. The inclination of the orbit to the plane of the equator (48° for TIROS I–IV, 58° for TIROS V–VII) further reduced the scope of the satellites so that only 20% of the Earth's surface could be photographed each day.

TIROS IX, launched in January 1965, provided the remedy by flying in a near-polar orbit with two cameras pointing outward from the rim of the satellite in the orbital plane. The cameras were mounted 180° apart and were triggered only when pointing directly downward at the Earth's surface. Because the orbit was near-polar it could be made Sun-synchronous, that is, it maintained the same position with respect to the Sun while the Earth rotated beneath it (see Section 4.1). This enabled the satellite to provide the first complete daily coverage of the entire Sun-illuminated portion of the globe. By early 1965 the TIROS pictures were sufficiently reliable to form a complete image of virtually the whole global atmosphere.

At about the same time the more ambitious Nimbus programme was beginning to produce results. These satellites were triple the weight and size of the TIROS satellites and were used to test new cameras and sensors. Nimbus 1 was launched in August 1964 and Nimbus 2 in May 1966. These satellites also had a more advanced stabilisation system that enabled them to keep the cameras and other sensors pointing earthward at all times. They operated successfully and proved extremely effective for testing advanced instruments. A parallel advance was the introduction of the TIROS Operational Satellite (TOS) system with the launch of three satellites during 1966. Because these were operated by the Environmental Science Services Administration (the forerunner of NOAA) they were designated ESSA 1, 2, and 3.

The quality of data from these satellites was sufficient to be integrated into routine cloud analyses. Digitised mosaics and hand-prepared global images were made in the United States and transmitted to many locations around the world. Apart from being used to track major storms these data had other interesting applications. When integrated over periods from 5 to 30 days they could be used to measure average cloudiness and to detect the extent of snow and ice cover. The latter data were soon being used to route ships taking supplies to research stations in Antarctica and to provide weekly maps of ice cover on the Great Lakes for cargo ships.

The ESSA series consisted of a total of nine satellites launched between 1966 and 1969. They were succeeded by the ITOS (Improved Tiros Observational System) series in 1970 which carried more advanced sensor systems. After ITOS 1, these satellites were designated NOAA 1–5, the last of which was launched in July 1976. The next stage was the much larger TIROS-N series which started with the launch of TIROS-N in October 1978 and was followed by NOAA 6, 7, 8, 9, 10 and 11. These satellites carried a range of increasingly advanced instrumentation. But to understand the historical development of this equipment we must go back to the experimental satellite programmes that were running alongside the TIROS and ESSA operational series.

3.3.2 *Geostationary satellites*

An important parallel work to polar-orbiting satellites was the development of geostationary satellites. At an early stage, it was realised that the solution to providing continuous coverage of the weather from space was to launch a satellite into an equatorial orbit so that it remained over the same spot continually (see Section 4.1). Following the successful demonstration of the geostationary communications satellite Early Bird in 1965, the first Earth-viewing system (NASA's Applications Technology Satellite, ATS 1) was launched into an equatorial orbit over the Pacific on 7 December 1966. This satellite, designed to test the applications of geostationary satellites, was not planned to include any meteorological equipment. Professor Verner Suomi of the University of Wisconsin realised, however, that a spinning geostationary satellite would offer an ideal system for taking global cloud pictures from which wind speed could be estimated. Only months before the launch of ATS 1 in 1966 he obtained permission from NASA to install a small scanning telescope on the satellite. This consisted of a 6-inch aperture telescope on gimbals. A tilting mechanism and a photomultiplier was mounted on the cylindrical structure of the spacecraft so as to scan the face of the Earth once on each revolution around the north–south spinning axis. Because of the action of the tilting mechanism, different lines were scanned on successive revolutions thus building, line-by-line, a full television image of the Earth.

In spite of its hurried preparation, the equipment was outstandingly successful. The jitter between successive pictures did not exceed 1 km and the instrument made meteorological observations for six years. These enabled measurements to be made of cloud movements and wind speeds over vast areas of the tropical Pacific where no observations had existed before. It was immediately apparent that a set of four or five such satellites dotted around the globe above the equator could provide an immensely valuable global observation system. This led to major efforts in both Japan and Europe to provide satellites to complement the two operational systems

Table 3.3. *Geostationary and experimental meteorological satellite programmes*

Geostationary

ATS 1 and ATS 3[a].   The two NASA Advanced Technology Satellites launched in 1966 and 1970, respectively, which demonstrated the capacity of geostationary satellites to make useful meteorological measurements.

SMS/GOES.   The US series of operational geostationary environmental observation satellites usually sited at longitudes of 75° W and 135° W. The first was launched in May 1974 and GOES 7 was launched in March 1987.

Meteosat series.   The European satellites sited over the Greenwich Meridian. Meteosat 1 was launched in November 1977 and failed in June 1979. Meteosat 2 was launched in June 1981 and the latest – Meteosat 4 – in March 1989.

GMS series.   Japanese satellites stationed over the Pacific at 140° E since June 1977.

Experimental satellites

Nimbus series.   NASA satellites used to test a wide variety of environmental measurement techniques, starting with launch of Nimbus 1 in August 1964 and concluding with launch of Nimbus 7 in October 1978.

GEOS 3.   One of a series of NASA satellites launched in 1975 and specifically designed to make a range of geophysical measurements.

SEASAT.   A satellite (launched in July 1978 and failed in October 1978) designed to test a variety of microwave systems for measuring the surface of the oceans.

GEOSAT.   A satellite launched in March 1985 and specifically designed to measure the height of the oceans' surfaces and the levels of the ice-caps (eg., of Antarctica and Greenland).

[a] ATS 2 failed on launch.

the US planned to install by the mid-1970s. The aim of these efforts was to have a complete set of geostationary satellites in place for the First GARP Global Experiment (FGGE), organised by the World Meteorological Organisation (WMO), in 1979, which was designed to pool ground-based and satellite observations in a coordinated way to evaluate the potential utility of all the data.

The programmes that led up to the achievement of the objectives of FGGE were spearheaded by the United States. Following the success of ATS 1 and ATS 3 (ATS 2 was lost on launch) NOAA launched the first in a new series of Earth-synchronous satellites in May 1974. Designated SMS/GOES 1 (Satellite Meteorological Service/ Geostationary Observational and Environmental Satellite 1), it obtained not only visible but infrared images. This equipment worked on the same principle as the earlier ATS instruments. The visible images had a ground resolution at nadir of 4 km for full-disc images and higher resolution images (either 2 km or 1 km) if it concentrated on a particular sector. The infrared images which could be obtained during day and night, because they measured terrestrial radiation in the 10.5–12.6 μm region, had a resolution of 9 km. With the launch of GOES 2 in 1975 NOAA had geostationary satellites sited over the equator at 135° W and 75° W so that almost all of North America, South America, and the adjacent ocean areas were under con-

observation. Full-disc images were routinely scheduled at 30-minute intervals. Alternatively, more limited north–south sectors during storm days were scheduled at shorter time intervals such as 15, 5, and 3 minutes.

The first Japanese satellite (GMS 1) was launched in June 1977. Stationed near 140° E and operated by the Japanese Weather Service, it was able to provide images of East Asia, much of the Pacific, and Australia. In November 1977 the European Space Agency's Meteosat was launched to take station close to the Greenwich Meridian over Africa and provide images of Africa and Europe. With the launch in 1978 of the third US satellite (GOES 3), which enabled GOES 1 to be moved to a new position over the Indian Ocean at 58° E, the Earth was ringed by five geo-stationary satellites which could transmit a continuous set of images of the entire globe between around 55° N and 55° S.

These satellites provided images in both visible and infrared regions. But these basic measurements were extended by new instruments. Meteosat contained an additional channel which operated at different infrared wavelengths (5.7–7.1 μm). This was able to detect the amount of water vapour in the atmosphere and so infer additional information about weather patterns at different levels (see Section 4.6). More comprehensive examination of the atmosphere at various levels became possible with the launch of a more sophisticated imagery system on GOES 4 in 1980. This was capable of not only making standard visible and infrared images but also of inferring the temperature structure of the atmosphere and other meteorologically important parameters such as the amount of water vapour at different levels. Similar instruments have been flown on GOES 5, 6, and 7, and are now used in a variety of forecast work.

3.3.3 Experimental programmes

Alongside the development of operational meteorological satellites there have been a variety of experimental systems which have not only tested new equipment but also made major contributions to both weather forecasting and wider understanding of the climate system. As noted in Section 3.3.1, the most important of these programmes was the NASA Nimbus series. The first of these (Nimbus 1) was launched in August 1964. But the truly important step forward in measurement techniques came with the launch of Nimbus 3 in April 1969. Nimbus 3 carried three different infrared sensor systems to measure atmospheric and surface temperatures. This method of temperature measurement is possible because the atmosphere and the surface beneath it emit heat radiation (see Section 2.1.3). Depending on whether the atmosphere absorbs infrared radiation strongly or weakly at a given wavelength, the detector will effectively observe upwelling radiation from a certain level. The amount of radiation measured is directly related to the temperature of the atmosphere. Early work in the TIROS series had demonstrated that rather crude maps could be made of atmospheric conditions from satellite data. The equipment on Nimbus 3 demonstrated, however, that this technique could make useful measurements of both the temperature of the atmosphere at different levels and the surface below, at any time of day.

The success of the infrared techniques was immediate and wide-ranging. Within six weeks of launching Nimbus 3, the temperature measurements became a regular input to the analyses by the US National Weather Service of the atmospheric circulation patterns over the Northern Hemisphere. While the observations from Nimbus 3 gave only a single temperature reading for a given level in the atmosphere over an area 225 km square, the global coverage was able to provide useful information about large-scale weather patterns. Although it would be many years before this type of measurement would have a significant impact on the quality of computer-generated forecasts, the use of infrared radiometers for temperature sounding became an integral part of satellite meteorology. The first operational radiometer designed to measure the temperature profile of the atmosphere was launched in NOAA 2 in 1972. Since then successive generations of infrared radiometers are providing better measurements, producing much higher spatial resolution.

Another major advance in satellite technology was in the introduction of microwave measurement techniques. Here the Russians were first on the scene. COSMOS 243 carried a microwave instrument which made observations of water vapour and liquid water in the atmosphere, but it only worked for two weeks. The first American microwave equipment was flown on Nimbus 5, launched in December 1972, and provided a better guide to the potential of such instruments.

The major benefit of using microwave systems is that it is possible to see through clouds, which are opaque to infrared radiation. The atmosphere emits tiny amounts of heat radiation in the microwave region. These wavelengths are long compared to the size of the droplets in clouds and so microwaves are virtually unaffected by cloud cover. Thus microwave radiometers can cover areas where infrared measurements cannot probe. Furthermore, because the amount of microwave energy emitted varies with the nature of the surface, these techniques are particularly effective for measuring such features as sea ice, snow cover, and soil moisture. They can also be used to infer wind speeds from the surface roughness of the oceans.

The measurement of microwaves welling up from the Earth is only part of the story. Because microwave devices can generate radiation efficiently without large power requirements, satellite instruments are not restricted to a passive role of measuring the Earth's microwave radiation. They can also send out their own signals and measure what is scattered back to receivers on the spacecraft. This provides information about the Earth's surface properties. These techniques were first put into practice in a Skylab experiment in 1972 and then with the launch of a satellite designed to undertake geophysical studies (GEOS 3) in 1975, and SEASAT in 1978. The latter was a test bed for a variety of active microwave instruments. The most

Table 3.4. *Growth of sensor technology on Nimbus satellites* (*showing details of the different spectral regions used in developing new measurement techniques*)

	Nimbus series						
	1	2	3	4	5	6	7
Launch date (month/year)	8/64	4/66	4/69	2/70	12/72	4/75	10/78
Number of experiments	3	4	9	9	6	9	9
Number of spectral channels	3	8	28	43	34	62	79
Spectral regions[a]							
Visible	X	X	X	X	X	X	X
Near-infrared	X	X	X	X	X	X	X
Mid- and far-infrared			X	X	X	X	X
Ultraviolet			X	X		X	X
Microwave					X	X	X

[a] X indicates where instruments working in the given spectral regions were installed on the various satellites in the Nimbus series.

simple of these systems was a radar altimeter for measuring the height of the surface of the sea. From this measurements of ocean currents, winds, and waves could be made. It also measured the height of the polar ice-caps. A more complicated piece of equipment was a radar scatterometer which could measure wind speed and direction.

The most sophisticated microwave instrument developed to date is a synthetic aperture radar (SAR) system which was capable of taking pictures of the surface roughness of the Earth. The equipment flown on both SEASAT and on the Space Shuttle has shown that it is possible to make detailed images of the wave patterns beneath hurricanes or the structure of pack ice in polar regions. While this technique has important applications far beyond meteorology, extending through a range of earth sciences, such as oceanography, glaciology and geology, it offers the prospect of a fascinating insight into the dynamic properties of weather systems. Although it only operated for three months, SEASAT revolutionised the perceptions of meteorologists and oceanographers and provided large amounts of data. More importantly, its success provided the basis for the next generation of oceanographic satellites which will come into service in the next decade.

Alongside these principal measurement techniques the experimental programmes have explored a wide variety of alternative ways of probing the atmosphere. In particular, the Nimbus series carried a rapidly expanding range of sensors (see Table 3.4) which explored different aspects of the radiation exchange properties of the atmosphere and the surface beneath. Culminating in the immensely successful Nimbus 7, which operated for over eight years, this series extended the sensitivity of visible, infrared, and microwave equipment, and also developed other technologies such as the investigation of scattered solar ultraviolet radiation. This radiation is modified by the presence of trace constituents in the atmosphere, notably ozone. This gas, which is an important trace constituent in the stratosphere, shows substantial seasonal variations. Such variations are of meteorological importance in that they influence the behaviour of the upper atmosphere, and are also of environmental interest because of their possible link with man-made pollutants.

Some measurements of ozone using both infrared and ultraviolet equipment were made on Nimbus 3 and 4. But, the ability to measure ozone in the upper atmosphere on a routine basis, achieved with the instrument to detect scattered solar ultraviolet radiation on Nimbus 7, was a major environmental breakthrough. It produced accurate global maps of the total amount of ozone and its vertical distribution in the upper atmosphere. Such maps have played a central role in studying the decline of ozone levels over Antarctica. Equipment on Nimbus 7 and the Space Shuttle has also demonstrated that it is possible to detect man-made pollutants such as sulphur dioxide, carbon monoxide, and methane. While the immediate impact of these trace constituents on the weather is negligible their potential long-term involvement in climatic change may prove to be one of the most important meteorological issues of the coming years.

4

Satellite instrumentation

'We see things not as they are, but as we are.'
H. M. Tomlinson

THE INSTRUMENTS flown on modern weather satellites are the result of a lengthy evolutionary process. They have been developed from remote-sensing and image-forming devices which were already widely used in other fields (eg., television cameras, infrared radiometers, and radar). In this chapter we will consider how such ground-based equipment has been adapted and refined to meet the needs of satellite meteorology. But before doing so we need to consider the basic features of satellites and their orbits.

4.1 Satellites and their orbits

Weather satellites come in many shapes and sizes. Many of the mechanical features are specific to individual satellites and do not need to be examined in detail. What is important is to identify those features of satellites which are essential to the effective operation of satellite systems. These are defined by three requirements: the first is that one axis of the satellite must be kept fixed in space so that the instruments can produce reproducible observations of the Earth; the second is an adequate and reliable power supply to maintain the continual operation of the on-board instrumentation; and the third is the orbit of the spacecraft.

The stability of weather satellites has relied on the gyroscopic principle. Either the whole spacecraft spins or alternatively a circular momentum wheel spins continually. In addition occasional fine adjustments of the attitude of the spacecraft can be made by small gas jets using an on-board supply of gas – usually hydrazine. The sources of power are arrays of solar cells which convert sunlight directly into electrical energy. But the amount of power is limited and so the instruments have to be economical with power supplies. When this requirement is combined with the fact that the weight of all equipment is severely restricted, it is evident that all instruments have to be designed and tested with great care to provide the right combination of efficiency, lightness, and reliability.

There are two forms of orbit used by weather satellites and these define what can be measured. Because the Earth is viewed in different ways from each of these orbits, instrumentation has to be considered in terms of the fundamental properties of them. Any orbit can be regarded as a circular ring about the centre of the Earth. Effectively it is fixed in space as the Earth rotates beneath it. If the orbit passes over the poles the Earth will spin under it every 24 h, so that any point on the surface will pass below the orbit every 12 h. A satellite on such a polar orbit will pass over the same place at the same time of day. But, because the Earth orbits the Sun, this time of day will change by 24 h through the year. To keep the orbit truly synchronous with the time of day, it must be offset by a small angle to the polar axis. Then the satellite is subjected asymmetrically to the non-uniform gravitational field of the Earth, because it is not exactly spherical and this causes the orbit to precess. With the right choice of angle, this leads to a 'Sun-synchronous orbit' in which the Earth passes under the satellite at the same times by the Sun each day.

A satellite in a Sun-synchronous orbit at an altitude of about 850 km takes about 100 minutes to encircle the Earth. If the satellite broadcasts the observations it is making instantaneously these can be picked up by ground stations. For any ground station a satellite will be above the horizon when it is within about 3000 km of it. This means that from a single spacecraft a ground station can collect images from within a 3000 km area every 12 h.

The 'geosynchronous orbit' is a more simple concept. If a satellite is put into an equatorial orbit, it will continue to circle the Earth above the equator. If the orbit radius is chosen at a certain value, then the period of the orbit can be exactly 24 h which means that the Earth will rotate beneath at precisely the same angular velocity. So the satellite will remain above the same point over the equator. This is achieved when the satellite is about 35 900 km above the equator.

From such a position the satellite can view the face of the Earth between about 55° N and 55° S latitudes and to 55° of longitude on either side of its position over the equator. But toward the edges of the field of view the obliqueness causes increasing loss of detail, especially beyond latitudes 50° N and S. This loss of detail is less of a problem at the longitudinal edge of view as there is the possibility of overlapping the images from adjacent geostationary satellites. But for most purposes meteorologists rely on a single satellite which covers their area of interest. Because the satellite can see the entire field of view all the time, it is natural that instruments on the platform should be designed to exploit this vantage point as fully as possible by producing frequent pictures of the whole visible disc.

One other feature of satellites must also be emphasised. All observations made by weather satellite are transmitted back to Earth by radio. To form images of what is going on below, the scene is scanned in a sequential manner. The measurements of each element in this scanning process are then transmitted to a ground receiver and the image recreated from these data. The most obvious way to do this is to use a television camera which projects the scene it is viewing onto a photoelectric screen. This screen is scanned by an electron beam to produce an electric current. The instantaneous magnitude of this current is proportional to the brightness of the portion of the screen being scanned. This electric signal can then be converted into a radio signal and transmitted over great distances.

The early TIROS satellites used forms of television cameras to obtain the first pictures of the Earth's weather. But, from the outset, it was an objective to make quantitative measurements of both the reflected sunlight and upwelling terrestrial radiation. The most effective way to do this is through a scanning radiometer, which measures the radiation from the Earth over a specified wavelength range.

4.2   Radiometers

A radiometer is an instrument which makes quantitative measurements of the amount of electromagnetic radiation incident on a unit area in a specified wavelength interval. Typically in a satellite system a radiometer consists of an optical system designed to view a small area of the Earth through an angled mirror. This mirror rotates at high speed so that the radiometer views successive swathes of the Earth below. These swathes are perpendicular to the flight path of the satellite and typically extend some 1500 km on either side of the orbit.

The radiation entering the radiometer is split up into a number of separate beams, each one of which passes through an optical filter to select a given wavelength region and is then measured by a separate detector. Depending on the design of the optics, the detectors can make measurements of either the successive circular or square areas in the swathe below as the mirror scans across the field of view. Each measurement is used to form a pixel in the image which is built up from this string of data. The first multichannel infrared radiometer was flown on TIROS VII.

The spatial resolution of the radiometers is defined by the size of the detectors and the optical parameters of the viewing system. A number of detectors can make measurements at the same time at different wavelengths. Each detector is chosen for its sensitivity at a given wavelength, and the narrow band of wavelengths to be studied is usually selected using some form of optical filter, although other means of spectroscopic isolation of the desired wavelengths have been tested with considerable success. Modern optical interference filters can isolate a band about 3 cm^{-1} wide at a frequency of 667 cm^{-1} (15 μm). The spatial resolution is chosen on the basis of the sensitivity of the detectors and the required signal-to-noise ratio for each observation – higher sensitivity or lower acceptable signal-to-noise ratio permits the radiometer to make measurements of smaller areas.

Because radiometers must make absolute measurements of the amount of radiation falling on their detectors, they must be calibrated continuously. This is done by viewing one or more standard on-board sources of known temperature when the rotating mirror is pointing away from the Earth. Typically these are cavities which act effectively as black bodies at a temperature around 288 K, although a cavity of lower temperature may also be used. The radiometer also views deep space which has a very low radiative temperature, and this provides a near-zero radiation signal. These measurements provide calibration points against which to check the equivalent temperatures of the observations made by the radiometer when viewing the Earth.

Radiometric observations of terrestrial radiation can also be made in the microwave wavelength range. Because the amount of upwelling microwave radiation is very small the measurement techniques have to be correspondingly more sensitive. This sensitivity is achieved by using radio detection methods. Known as superheterodyne amplification, this technique makes use of an on-board source of coherent microwave radiation to greatly strengthen the incoming signal. The benefit

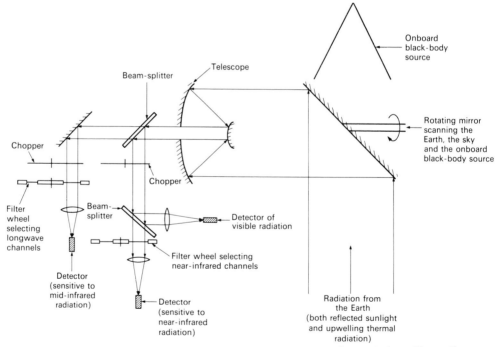

Fig. 4.1. This schematic diagram shows the principal components of a typical satellite radiometer showing how both reflected solar radiation and terrestrial radiation are split into various wavelength bands to enable measurements to be made of the properties of the Earth's atmosphere and surface.

of this approach is that not only is it possible to observe very faint signals, but the frequency of detection is precisely defined in relation to the frequency of the on-board microwave oscillator. So the frequency can be chosen to coincide precisely with the required absorption coefficient in the spectrum of either oxygen or water vapour lines to allow the radiometer to probe the temperature at a given level. But this increased selectivity is counterbalanced by the much smaller change in the radiated energy for a given change in temperature (see Section 2.1.3).

4.3   Radiometric measurement of the atmosphere and the Earth

To consider how satellite radiometers measure the properties of the atmosphere and the Earth's surface we need to extend the discussion in Chapter 2. The radiative processes of the atmosphere presented there define what can be observed from space. In addition the vertical structure of the atmosphere plays a crucial part in efforts to measure the temperature and humidity at different levels.

4.3.1   *Vertical distribution of pressure and temperature*

The vertical structure of the atmosphere also plays an essential part in defining the amount of outgoing terrestrial radiation. Atmospheric pressure always decreases with height, but there is no simple pressure–height relationship because this depends on atmospheric composition and temperature. As a simple rule of thumb pressure falls to about a half the surface value at an altitude of 5 km and continues to decrease very roughly by a factor of two in each successive 5 km interval. This means that three-quarters of the atmosphere is contained below an altitude of 10 km.

Temperature also varies with altitude, but its behaviour is more complicated. In the lower atmosphere the average rate of decrease upward is about $6\,°C\,km^{-1}$. This vertical gradient is commonly referred to as the 'normal lapse rate'. It varies with height, season, latitude, and other factors. The normal lapse rate extends through the lower atmosphere, often known as the 'troposphere', where convection dominates energy transfer to a height which varies from around 20 km at the equator to < 10 km at high latitudes. Above the troposphere there is a region where temperature is roughly constant, and so the lapse rate is zero. At these levels energy transfer is dominated by radiative processes and there is no appreciable vertical mixing. For this reason, these levels are known as the 'stratosphere'. The boundary between the troposphere and the stratosphere is defined as the 'tropopause'. At greater altitudes the temperature rises. We will focus on the troposphere, where virtually all the weather occurs. Only in limited areas will we look at changes in the stratosphere which could influence the weather or lead to climatic change.

The temperature structure of the lower atmosphere can be much more complicated than the standard description given above. In particular, the temperature profile can be inverted at the surface and sometimes at higher levels. Such inversions are most common where there is strong radiative cooling of the surface, for instance, at night in calm conditions and in the Arctic during the winter. They can also be the result of the interaction of air masses of different origins in dynamic weather systems. This type of temperature profile can be important in creating the right conditions for exceptionally low temperatures at ground level or for trapping pollutants near the surface.

Another important aspect of the temperature profile is the effect of water vapour. When moist air rises and cools water vapour eventually condenses to form clouds. In so doing it releases the energy that has been used to convert liquid water into water vapour (the 'latent heat of vaporisation'), and this warms the air. This increases the buoyancy of the air, and depending on the lapse rate may lead to it rising still further and leading to yet more condensation. When the air is warm and humid and the lapse rate is high this can produce towering clouds which lead to severe

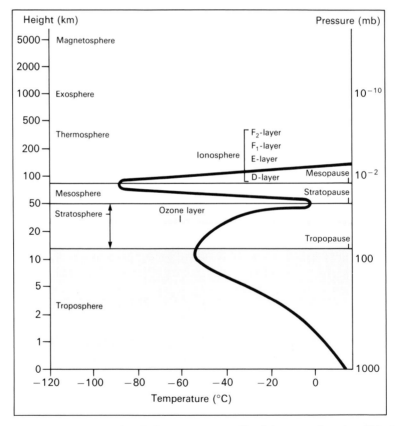

Fig. 4.2. The vertical structure and typical temperature profile of the atmosphere in mid-latitudes. For most purposes active weather systems can be regarded as being restricted to the troposphere.

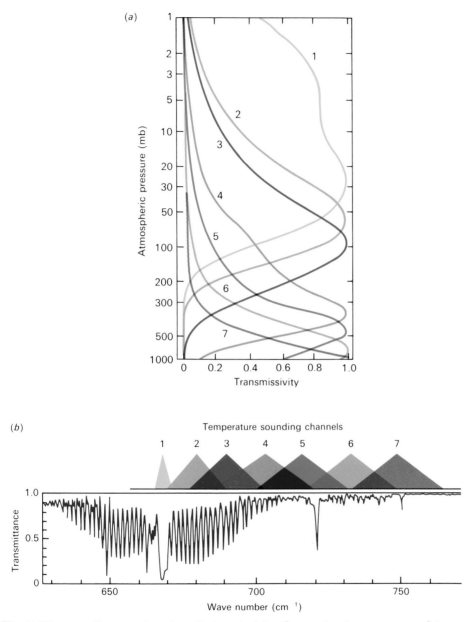

Fig. 4.3. These two diagrams show the radiative principles of measuring the temperature of the atmosphere. Terrestrial radiation emitted in wavelength regions where the atmosphere is largely transparent will come largely from the lower levels of the atmosphere or from the surface, while radiation emitted in wavelength regions where the atmosphere is opaque will come principally from the upper levels of the atmosphere. (a) The normalised weighting functions for the radiance received from different levels in the atmosphere by each of the seven temperature-sounding channels. (b) High resolution spectrum of the transmittance of the Earth's upper atmosphere in the mid-infrared sounding channels of the High Resolution Infrared Sounder of the TOVS system in the TIROS-N/NOAA satellites.

47

storms. The combination of humidity and temperature profiles which create such unstable conditions are of great interest to meteorologists trying to predict such storms.

The pressure and temperature profiles of the atmosphere exert major influences on the form of terrestrial radiation emitted to space. The dominant effect is that the number of molecules of carbon dioxide is directly related to pressure, so the amount of infrared radiation absorbed and re-radiated by carbon dioxide falls off with increasing altitude in proportion to the pressure. The behaviour of important absorbing species such as water vapour and ozone is more complicated. The number of water moelcules falls off with altitude even more sharply than carbon dioxide because of the condensation effects noted above. Ozone is different. Because it is created by the action of sunlight high in the stratosphere its concentration peaks at altitudes of around 25 km, so the radiative effects of ozone relate to these levels of the atmosphere.

Because the amount of energy radiated by the trace gases is related to their temperature, the temperature profile of the atmosphere exerts a strong influence on the radiation emitted to space. In principle, this means that the lower warmer levels of the atmosphere will dominate the radiation patterns. This basic premise is, however, complicated by two additional factors: first, the radiation emitted by, say, carbon dioxide in the lower levels of the atmosphere will be absorbed and re-radiated by colder carbon dioxide in the upper atmosphere; the second factor is the effect of the major atmospheric constituents, nitrogen and oxygen. While these gases do not make a significant direct contribution to the absorption and emission of infrared radiation they do alter the radiative properties of the trace gases. This alteration is caused by intermolecular collisions which have the effect of broadening the spectral features of radiatively-active gases (see Section 2.3.1). Since this process is proportional to total atmospheric pressure it means that the spectral features are broader at low levels and become much narrower at high altitudes. The result is that the radiation escaping to space will effectively come from the upper atmosphere if its wavelength is close to the centre of a spectral feature and from lower levels if its wavelength is well away from the centre of the spectral feature. With suitable instrumentation this effect can be exploited to discriminate between radiation coming from different levels in the atmosphere.

4.3.2 *Temperature sounding of the atmosphere*

The principle of measuring the temperature of the atmosphere from above relies on using an infrared or microwave radiometer to measure the upwelling terrestrial radiation at different wavelengths. This radiation consists primarily of emissions from the surface attenuated by the atmosphere, plus radiation emitted by the atmosphere attenuated by the levels above. Over partially reflecting ('non-black') surfaces, both downwelling atmospheric and solar radiation are reflected from the surface and after attenuation by the atmosphere can also contribute to radiation in some circumstances.

Absorption and emission due to carbon dioxide are most frequently used to sense the temperature of the atmosphere. The advantage of carbon dioxide is that it is mixed uniformly in the atmosphere in known quantities, whereas concentrations of the other molecules that absorb strongly in the infrared (eg., water vapour and ozone) vary greatly over time and space. Carbon dioxide has a number of infrared absorption bands due to the fact that at certain frequencies the molecules can be excited to vibrate and rotate by absorbing electromagnetic radiation. This process occurs at infrared wavelengths of around 4.3 μm and most importantly from 12 to 18 μm, centred on 15 μm. These absorption bands have a complicated fine structure due to the rotational motion of the molecule. This means that the absorption co-efficient of carbon dioxide can vary by many orders of magnitude over a short wavelength interval. This property can be used by satellites to peer down to specific depths in the atmosphere.

The amount of upwelling radiation can be calculated from known properties of the surface and absorptive behaviour of the atmosphere at any given wavelength. This involves a detailed mathematical analysis, often termed an 'inversion tech-nique', which integrates the combined effects of emissivity and absorptivity of each level of the atmosphere together with those of the surface below. Such calculations show that where the atmosphere is strongly absorbent most of the upwelling radi-ation comes from the highest levels, whereas at wavelengths where the atmosphere is virtually transparent radiation comes from close to the surface. By choosing to operate at a given wavelength (and hence absorption coefficient), it is possible to sense the temperature at different levels in the atmosphere.

A source of error is introduced in the method of calculating the temperature pro-files. There are an infinite number of atmospheric temperature profiles which could, in theory, produce the observed radiances. So to get near to the correct result the mathematical inversion technique must make a 'first guess' to get started and then refines. Normally, this initial estimate uses the climatological average for the time of year and geographical position. This works well if the weather is near normal, but becomes increasingly inaccurate in more abnormal conditions. In particular, if the tropopause is anomalously low or high there are problems. One way round this difficulty in forecasting work is to use the forecast profile for the time of observa-tion as the 'first guess' (see Chapter 12).

Because satellite-borne radiometers sense the integrated radiative effect of the whole atmosphere, a measurement of upwelling energy at any given wavelength does not relate to the temperature at a single level. Instead, it provides information about the temperature of a slice some kilometres thick centred on a given level. To obtain a vertical temperature profile of the whole atmosphere a number of infrared wavelengths are used (see Table 4.2). Under clear conditions this can provide measurements to an accuracy of better than 1 K. But in practice, aerosols and clouds in the field of view mean that on average temperature measurements have a mean error of about 2 K above an altitude of 4 km and somewhat higher at lower levels. Moreover, because each measurement is covering a broad altitude range in practice, for work such as numerical weather forecasting independent figures can only be obtained for four levels.

By measuring in the microwave region, the problem of absorption by aerosols and clouds is eliminated since the wavelength is much greater than the size of aerosol and cloud particles. Although the amount of heat radiation emitted at these wavelengths is much smaller than in the infrared, more sensitive detection techniques are available. Moreover, unlike in the infrared, oxygen has a set of absorption features at these wavelengths. Using the absorption band in oxygen at wavelengths around 5 mm, it is possible to make the same type of temperature measurements as in the infrared.

4.3.3   *Surface temperature measurements*

In the transparent 'window' regions well away from the absorption bands of atmospheric gases (eg., 10.5–12.5 μm) most radiation comes from the Earth's surface. Providing we know what the emissivities of various Earth surfaces are, then it is possible to obtain accurate measurements of surface temperature. But the wide variety of surface conditions requires the calibration of both infrared and microwave radiometers to give reliable results. In the infrared this calibration is relatively simple as it has been found that emissivity is relatively insensitive to surface conditions, and values of 0.85 over land and 0.96 over the oceans are used. In the microwave region it is more complicated. The emissivity of the open ocean is typically 0.45–0.65, increasing with decreasing temperature and increasing wind speed. The emissivity of land is uniformly high being between 0.90 and 0.95, but it is affected by soil moisture. Sea ice has an emissivity of the order of 0.7 or more, but it is altered by the amount of salt present which changes with the age of the ice (see Chapter 11). Snow has an emissivity of 0.90 or less, depending on both the wavelength and the depth.

The fact that the emissivity of the Earth's surface in the infrared is relatively uniform does not remove all the problems of surface temperature measurements. As with observing atmospheric temperatures, interference due to aerosols and clouds occurs. So, although it is possible to sense the upwelling heat radiation from the surface in the most transparent 'window' regions, at any one time much of the globe remains hidden from view. The way around this problem is to use microwave radiometers. However, accurate measurements of emissivity at microwave frequencies is complex. Microwave emissions over the oceans are very complex but also potentially the most useful. This is because emissivity is a function of surface roughness, and hence wind speed. So, in theory, microwave measurements of apparent water temperature can provide information not only about the actual temperature of the water, but also the wind speed above the water. But this requires knowledge of both how emissitivy is related to the surface roughness and also how this roughness is linked to the local wind speed.

At the lowest wind speeds the movement of the air produces ripples with a wavelength of a few centimetres. As the wind increases it starts to generate the much longer waves that are the familiar measure of the sea-state. The amplitude of these waves is initially proportional to the wind speed, but they do not die down quickly even when the local wind abates. At high wind speeds there is an increasing tendency for 'white caps' to form. All of these effects influence the amount of microwave energy radiated from the ocean surface. However, with exhaustive analysis it has been possible to unscramble the temperature and wind effects and demonstrate that accurate wind measurements can be made using satellite microwave radiometers (see Chapter 10).

The values of the microwave emissivity of snow and ice are sufficiently different from those of the oceans and dry land to make measurements of their extent. Depending on the microwave frequency used the emissivity of sea ice can be up to double that of the surrounding oceans, so the extent of pack ice can readily be observed using microwave radiometers. Similarly, the emissivity of snow is between 10 and 25% less than snow-free ground in the frequency range 10–40 GHz. Emissivity also decreases with snow depth. Furthermore, the amount of energy emitted depends on the proportion of liquid water in the snow. All of the above effects can be detected from space, and with careful calibration quantitative measurements of surface conditions can be made.

4.3.4   *Measuring water vapour*

Water vapour has a number of strong absorption bands in the infrared. The most important of these are at wavelengths around 6.7 μm and from 20 to 1000 μm. These can be used to make measurements from space. This is because the amount of water vapour in any spot in the lower atmosphere is likely to change appreciably with passing weather conditions and with the seasons, so the amount of upwelling radiation in, for example, the 6.7 μm region can be used to make direct measurements of the amount of water vapour at different levels in the atmosphere.

This technique relies on the same radiometric principles as temperature sensors. By choosing a wavelength where absorption, and hence emission, by water vapour is either strong or weak it is possible to infer the amount of water vapour present in the upper or lower atmosphere. Typically, satellite radiometers are fitted with channels that can measure water vapour in the lower, middle, and upper troposphere. These can provide important information about the motion and stability of the atmosphere which is of value in measuring wind and in forecasting the development of severe thunderstorms.

4.3.5 Cloud temperatures

Knowledge of cloud temperature is of considerable interest to meteorologists. Because clouds absorb so strongly in the infrared, radiometer measurements effectively provide the temperature of the top of thick clouds. So the temperature and height of cloud tops are generally obtained using the infrared 'window' channel in the standard scanning radiometer. Normally the height of the cloud top is then inferred from the typical temperature profile for the locality and time of year. Where the temperature profile of the atmosphere in the vicinity of weather systems is obtained using infrared sounders, more detailed analysis is possible. This is an interactive process. First, the presence of clouds is ascertained using infrared images. This information is then used to correct temperature profile measurements. The temperature of the cloud tops can then be matched more precisely with the temperature of the levels in the adjacent clear atmosphere to give a clearer picture of cloud structure. Such information is useful for making forecasts, as the deeper the clouds the greater their chance of producing severe weather conditions. Where there are thin high clouds it is more difficult to make measurements as emissivity is low and effective temperature does not correspond to cloud level but to a combination of the cloud and atmosphere below.

4.3.6 Scattered solar radiation

So far the sunlight reflected or scattered by the atmosphere, clouds, or the Earth's surface has been largely disregarded. For the most part this is reasonable because although many of the images produced by weather satellites have observed this radiation, they have simply recorded total flux over wide wavelength bands. This means that apart from the most obvious differences, such as variation in reflectivity of clouds, snow, and ice, outgoing solar radiation is not subjected to detailed spectroscopic analysis. But, as has been noted earlier, the atmosphere does absorb solar radiation in both the ultraviolet and the near-infrared. This means that if the spectrum of the scattered solar radiation is analysed it is possible to measure the concentration of trace constituents.

In addition the reflectivity of the Earth's surface has spectral characteristics which provide information about its properties. In the visible and near-infrared range this can be measured by comparing reflected radiation in different parts of this spectral range. In particular it is possible to make accurate measurements of the vegetation cover of the land surface and to assess the biological activity of the oceans. This can be done because chlorophyll absorbs strongly in the green portion of the visible spectrum, so by measuring the amount of sunlight reflected in this region as compared with that reflected in adjacent spectral regions, it is possible to measure the amount of photosynthetic activity. In the case of terrestrial vegetation this is best done by comparing the amount of solar radiation reflected in the visible region with that reflected in the near-infrared (0.75–1.1 μm). Because reflection is high in the infrared the difference is a reliable measure of the amount of vegetation present. In the oceans the combination of the effects of the water and biological activity produce more subtle variations in colour, and hence reflectivity, in the visible region, so making measurements in various parts of the visible spectrum provides a sensitive measurement of biological activity.

Another area where scattered solar radiation is used to make measurements is in the ultraviolet. In particular, instruments have been developed to measure the amount of ozone in the upper atmosphere. The most successful example of such equipment is the solar backscatter ultraviolet spectrometer (SBUV) on Nimbus 7. This measures radiation in the 0.34–0.26 μm range in 12 narrow spectral intervals. The four longest wavelength intervals are used to measure the total amount of ozone and the remainder to measure concentrations of the gas from altitudes of around 50 km down to roughly 22 km. The sensitivity of the ozone profile measurements depends on a number of factors, including the solar zenith angle, the total amount of ozone, and its distribution with height.

4.4   Experimental infrared radiometers

Modern infrared radiometers are now increasingly produced to a standard design. But in the process of developing these instruments a variety of approaches to making measurements was explored. Each involved a lengthy process. First, prototype systems were built and tested in the laboratory. Once the principle of the measurement technique had been established an experimental system was built and flown in a satellite, and the satellite data were compared with surface-based observations. Only when the satellite equipment had been satisfactorily calibrated could it be used for operational purposes. This laborious process is one reason why satellite systems took so long to develop and integrate in meteorological operations. The first scanning infrared radiometer which could measure the temperature of the atmosphere at different levels was flown in the Nimbus 3 satellite in 1969. It had taken ten years to develop. Designated the Satellite Infrared Spectrometer (SIRS-A), this instrument used a diffraction grating spectrometer to isolate eight spectral intervals. Each interval was 5 cm^{-1} wide; one in the 'window' region at 11 μm, and seven in the 13.3–14.9 μm region in which CO_2 emits radiation. The instrument was not designed to provide measurements with the spatial distribution and resolution needed for weather analysis but to demonstrate the capacity to make temperature soundings. It viewed a strip 225 km wide beneath the orbital track and took readings every 225 km. As a result its soundings were about 2700 km apart at the equator and, because of the large field of view, were frequently contaminated by the presence of clouds.

In spite of its limitations, SIRS-A was extremely successful in showing that accurate measurements of atmospheric temperature could be made from space. In clear conditions it was capable of making temperature measurements of the atmosphere, cloud tops, or the Earth's surface with an accuracy of just over $\pm 1\,°C$ at altitudes above 6 km and $\pm 2\,°C$ at lower levels. It was acclaimed as the most significant development in the history of meteorology. For the first time meteorologists were able to have a truly global measure of the temperature of the atmosphere where previously there had been huge gaps over the oceans and irregular cover over the continents. Even though data from this instrument were soon being used in forecast work with demonstrable improvements in performance, it was not until the 1980s that there was unequivocal evidence that radiometric temperature measurements from satellites were making a significant contribution to numerical weather forecasting.

The reasons for the slow progress were manifold but there were two in particular. First, the scale of the data flow initially exceeded meteorologists' capacity to use it. SIRS-A was capable of making 8000–10 000 soundings every 24 h. But, at first the US National Weather Service forecasting operation only used 400 of these. Secondly, while the instrument performed excellently, many of the figures extracted from the satellite observations did not match surface observations. In particular, it was found that if there were thin cirrus clouds, dust, or aerosols in the field of view, observed temperatures of the atmosphere or sea surface did not coincide with surface measurements.

Also the technique had difficulty in dealing with atmospheric temperature profiles which were markedly different from the climatological normal (see Section 4.3.2). Moreover, unexpectedly strong absorption in the humid, lower tropical atmosphere could distort the results. So a great deal of work had to be done before the full potential of infrared radiometers could be exploited.

The next stage in the development of infrared radiometers was SIRS-B, a smaller, lighter, more sophisticated descendant of SIRS-A, which was launched aboard Nimbus 4. This device sensed Earth radiances in 14 spectral intervals, eight of which were similar to those on SIRS-A and the remainder which measured emissions from atmospheric water vapour in the 18–37 μm region. The additional channels provided data from which gross vertical water vapour distribution could be calculated. The instrument was step-scanned so that it could look sideways to make soundings between, as well as along, the orbital tracks thereby improving geographic coverage. But its spatial resolution was 700 km. In spite of the success of both the SIRS instruments the designs were not adopted for operational systems, but they did serve as the initial test bed for temperature sounding.

The satellites Nimbus 3 and 4 were also used for testing other infrared radiometers. The second experimental system on Nimbus 3 was in some ways even more interesting. Called the Infrared Interferometer Spectrometer (IRIS), it was capable of measuring infrared terrestrial radiation over the range of 400–2000 cm^{-1} (25–4 μm) with a spectral resolution of 5 cm^{-1}. This enabled it to examine the emission properties of the atmosphere and the Earth's surface in greater detail than other spectrometers. Unlike other spectrometers which split the radiation into spectral components before measuring its intensity, IRIS detected all the radiation at once. The spectral characteristics were resolved by splitting the radiation into two equal beams, one of which travelled to a fixed mirror and back again while the other fell on a mirror which moved back and forth over a distance of a millimetre. When the beams were recombined they produced an interference effect which was recorded as a function of the mirror movement. This interference pattern could be measured and then mathematically transformed to calculate the spectrum of the radiation reaching the detector.

The IRIS on Nimbus 3 had a field of view of about 160 km across and made a measurement in 10 s. An improved version on Nimbus 4 covered a slightly reduced range of 400–1500 cm^{-1} (25–6.7 μm) but had a higher spectral resolution of 2.8 cm^{-1} and a smaller field of view of some 100 km. Both instruments were highly successful in making measurements of water vapour and carbon dioxide emission bands (and hence the temperature profile of the atmosphere) and also other atmospheric gases, notably ozone and methane. They also provided interesting information about the radiative behaviour of the Earth's surface; they showed that solar infrared radiation scattered by the sand of the Sahara desert could lead to misleadingly high surface temperatures being inferred at certain wavelengths.

The third experimental system flown on Nimbus 4 was the selective chopper radiometer. Designed and built at Oxford University, England, it exploited the subtle technique of using CO_2-filled cells as filters in conjunction with standard interference filters. The interference filters provided the coarse resolution with a bandwidth of between 3.5 and 10 cm^{-1}, while the CO_2-cells isolated the desired radiation within this band. The instrument had six channels. Four used cells filled with carbon dioxide at different pressures to select wavelengths at which the absorption coefficient of carbon dioxide was relatively constant within each channel. This improved the definition of radiation coming from different levels in the troposphere and lower stratosphere. The other two channels used carbon dioxide at lower pressure to 'selectively chop' the radiation coming from higher altitudes in the stratosphere. This was done by rotating two cells in front of the detector – one containing CO_2 and one not. This meant that the difference between the two signals measured the amount of radiation coming from the centre of the CO_2 lines. The advantage of this technique was two-fold. First, because the lines in the CO_2 spectrum at low pressure are much narrower than can be resolved by normal spectroscopic techniques (typically

0.01 cm^{-1}) it was possible to isolate radiation coming from higher in the atmosphere. Secondly, the method allowed the radiation from a number of lines to be measured, and hence enough energy could be detected to obtain adequate radiometric accuracy to make reliable temperature measurements in the upper atmosphere.

All three of these systems made major contributions to the development of modern satellite equipment. But in the case of SIRS and IRIS more functional optical filter systems were developed in parallel which were to result in the current generation of high resolution radiometers. In the case of the selective chopper radiometer, the technique proved to be the most sensitive means of measuring temperatures in the stratosphere. A modified version which relies on the pressure modulation of the gas cells to select the radiation emitted from the centre of the carbon dioxide lines is now incorporated into the operational equipment for current weather satellites.

4.5 The current TIROS-N/NOAA operational system

Experimental work of the Nimbus series together with the development of the operational experience of the first three generations of the TIROS series (TIROS, ESSA, and ITOS, see Chapter 3) have led to what has become the standard equipment for the current TIROS-N/NOAA series of satellites. The first of the sequence was TIROS-N launched in October 1978. It was followed by a series of NOAA satellites launched approximately every two years (see Table 3.2). Originally planned to consist of 11 satellites providing a global operational service during the period 1978–89, the series is now planned to extend beyond 1992. The principal components of these satellites are the Advanced Very High Resolution Radiometer (AVHRR), the TIROS Operational Vertical Sounder (TOVS), and the data collection system.

The AVHRR is the imaging device. Using the standard cross-track scanning mode, the initial models made observations in four spectral regions and current equipment has five channels (see Table 4.1). The visible and mid-infrared channels make the standard observations of reflected sunlight and terrestrial heat radiation. The additional channel in the near-infrared receives both reflected solar and terrestrial radiation, and so can be used to make different measurements by day and night. One valuable feature is that around $3.7 \ \mu\text{m}$ water droplets are highly reflective but ice crystals absorb radiation, so this channel can be used to discriminate between warm and cold clouds and make inferences about the freezing level.

The different wavelength bands are selected using a beam splitter to divide the incoming radiation into five components which then pass through optical filters to isolate the required bands. The spatial resolution immediately beneath the satellite is about 1 km in all channels. At the horizon this resolution is degraded to 2.5 km along the track and 7 km across the track. On each scan each channel samples 2048 discrete points. The equipment is extremely sensitive. In the visible channels the specified performance is a signal-to-noise ratio of 3 :1 for a 0.5% change in albedo.

Table 4.1. *TIROS-N/NOAA AVHRR channels*

Channel number	Wavelength (μm)
1	0.55–0.68
2	0.725–1.10
3	3.55–3.93
4	10.5–11.5
5	11.5–12.5

In the infrared the noise-equivalent temperature of the detector output is 0.12 K. All of the AVHRRs flown to date have exceeded these demanding specifications.

The TOVS consists of three instruments, providing temperature profiles of the atmosphere from the surface to an altitude of 32 km, tropospheric water vapour levels, and total ozone content. It includes:

(a) a high resolution infrared sounder, a 20-channel stepped scanned visible and infrared spectrometer (see Table 4.2) which was first tested on Nimbus 6. This is used to produce tropospheric and stratospheric temperature and moisture profiles with a spatial resolution of 20 km. As in the AVHRR the incoming radiation is split into the required number of beams and filtered optically before being measured.

(b) the stratospheric sounding unit furnished by the British Meteorological Office has three selectively chopped channels designed to measure temperatures around the 3, 8, and 20 mbar pressure levels. This is a pressure-modulated step-scanned system observing CO_2 radiances in the 15 μm region.

(c) the microwave sounding unit, a four-channel step-scanned spectrometer with response in the 5 mm oxygen band is used to obtain temperatures at the surface, two levels in the mid-troposphere (500 and 300 mbar) and the lower stratosphere (70 mbar). The spatial resolution beneath the satellite is a little over 100 km, declining to 300 km at the horizon.

The data collection system provides random access arrangements for the collection of meteorological, environmental, and scientific data from automatic platforms, both movable and fixed, such as buoys, balloons, and remote stations.

4.6 Geostationary satellite imaging systems

While the performance of equipment on geostationary satellites has advanced considerably over the past 20 years, the basic configuration has remained largely unchanged since the launch of ATS 1 in 1966 (see Table 3.3). The satellites have a cylindrical form. They are stabilised to spin about an axis parallel to the Earth's

Table 4.2. *High resolution infrared sounder channels*

Channel number	Wavelength (μm)	Frequency (cm^{-1})	Level of peak sensitivity (mbar)
1	14.96	668.40	30
2	14.72	672.20	60
3	14.47	691.10	100
4	14.21	703.60	280
5	13.95	716.10	475
6	13.65	732.40	725
7	13.36	748.30	Surface
8	11.14	897.70	Window sensitive to water vapour
9	9.73	1027.90	Window sensitive to ozone
10	8.22	1217.10	Lower tropospheric water vapour
11	7.33	1363.70	Middle tropospheric water vapour
12	6.74	1484.40	Upper tropospheric water vapour
13	4.57	2190.40	Surface (used for cloud correction)
14	4.52	2212.60	650
15	4.46	2240.10	340
16	4.39	2276.30	170
17	4.33	2310.70	15
18	3.98	2512.00	Window sensitive to solar radiation
19	3.74	2671.80	Window sensitive to solar radiation
20	0.69	14500.00	(Used for detecting clouds and correcting Channels 18 and 19 for reflected sunlight)

axis at a rate of 100 rpm. The outer walls of the cylinder are almost totally covered by solar panels which provide power for the on-board equipment. On the top of the satellite is an electronically de-spun aerial which points continuously toward Earth and transmits data to ground stations.

The Earth-viewing equipment is comprised of a reflecting telescopic system with an array of visible and infrared detectors at its focal plane. Depending on the properties of the optics and the size of the detectors, at any instant the system collects radiation from a small area on the Earth. Typically, the maximum resolution immediately below the satellite is 1–5 km in the visible and 5–10 km in the mid-infrared. Near the edges of the Earth's disc resolution is degraded by the angle of observation.

The viewing system scans the entire face of the Earth in sequence every half hour. The west to east scan results from the spinning of the satellite. During each revolution of about 0.6 s, the Earth is in view for only 0.03 s. In this short time the infrared detectors take 2500 consecutive observations of the upwelling radiation from the swathe 5–10 km wide scanned by the field of view of the telescope. At the same time two, or sometimes four, adjacent smaller visible light detectors scan correspondingly narrower swathes in the field of view and make twice, or four times, as many observations. During the remaining 0.57 s of a satellite rotation, after each scan, the telescope is advanced one step (approximately 1.25×10^4 radians, or 1/2500 of the Earth's disc) northward so that the next line scanned is adjacent to the previous line. In this way 2500 lines are scanned in about 25 min to give a complete image of the face of the Earth beneath the satellite. After the last line the telescope returns to its original position to be ready for its next half-hour scan.

The limitation of these basic Earth-viewing systems, which in the US geostationary satellites were known as Visible/Infrared Spin–Scan Radiometers (VISSR), is that they could not make temperature soundings of the atmosphere. To address this issue the first VISSR Atmospheric Sounder (VAS) was launched on GOES 4 in September 1980. This instrument, and subsequent models, worked on the same basic principle as the earlier VISSRs. The visible image system used a linear array of eight detectors to give a spatial resolution of 0.9 km. Temperature and moisture soundings were achieved using six thermal detectors to measure infrared radiation in 12 spectral bands in the region from around 4 to 15 μm. A filter wheel in front of the detectors was used to obtain spectral selection. Spatial resolution in the infrared was 7–14 km, depending on the detector used. Spectral bands were chosen to measure upwelling radiation from either CO_2 or water vapour at different levels in the atmosphere or from the Earth's surface in precisely the same way as in other atmospheric sounders.

Another important feature of the VAS is that it can be operated in three modes to meet different meteorological requirements. First, it operates in the simple VISSR mode, which provides a high resolution visible image and a single infrared image of the full Earth disc every half hour with a resolution of 7 km. The second is the multispectral imaging mode which produces a full disc image of atmospheric water vapour, temperature, and cloud distribution every half hour. It has the choice of doing this either with four spectral channels (the visible at 0.9 km resolution, the 11 μm window at 7 km resolution, and two others at 14 km resolution), or five channels (the visible plus any four other channels at 14 km resolution). Finally, the instrument can be programmed to dwell on a more limited area, using up to 12 spectral filters which can be positioned in sequence into the optical train while the scanner remains on a single scan line. The filter wheel can be programmed so that up to 255 scans are made on a single line in order to build up a more accurate temperature profile. The spatial resolution can be either 7 or 14 km. The VAS can also be instructed to cover a given number of scan lines to build up an image of a wider area.

The primary limitation of the VAS is that it cannot be used on its dwell mode at the same time as operating in the VISSR or multispectral modes. The best that has been achieved to date is to provide operational users with a 15-min VISSR image every half hour. In addition, parallel operation can produce 15 min of multispectral data and 11 min of dwell operation. So, special studies could be made of particularly interesting atmospheric developments.

4.7 Microwave radiometers

The standard design of satellite-borne microwave radiometers consists of a parabolic antenna which collects radiation and focusses it on to the detectors. The size of the antenna defines both the amount of radiation that is collected and the narrowness of the beam. The larger the antenna the more radiation can be collected from a smaller area. Current temperature sounding systems working in the oxygen absorption region (50–58 GHz) have antennae about 1 m in diameter and view an area approximately 100 km wide immediately beneath the satellite. They carry four detection channels to sense different levels of the atmosphere.

Another form of microwave radiometer works over a much wider frequency range so that it can sense water vapour, liquid water, precipitation in the atmosphere, and differences in the emissivity of the Earth's surface. The most advanced operational system of this type is the Scanning Multichannel Microwave Radiometer (SMMR) on Nimbus 7. This detects both vertical and parallel polarised radiation at 6.6, 10.7, 18, 21, and 37 GHz. The system consists of a single multifrequency

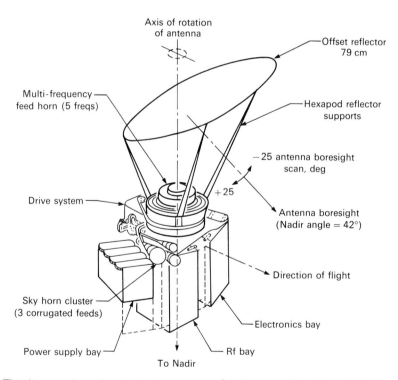

Fig. 4.4. This diagram shows the principal components of the Scanning Multichannel Microwave Radiometer (SMMR) which can make measurements of the radiative temperature of the Earth's surface in both clear and cloudy conditions.

horn at the focal point of an offset rotating elliptical parabolic reflector 1.1 m × 0.8 m. This has a conical forward-looking field of view which scans in azimuth at a constant angle of incidence to the Earth's surface up to 25° on either side of the satellite track, so it can cover a swathe about 800 km wide ahead of the satellite. The received beam width of the instrument is defined by the size of the parabolic reflector and the frequency, and varies from 4° at 6.6 GHz to 0.7° at 37 GHz. This provides a ground resolution of 100 km across-track and 150 km along-track at 6.6 GHz and 17 km × 27 km at 37 GHz.

4.8   Radar systems

Thus far, all the instruments described have measured either reflected sunlight or thermal radiation emanating from the Earth. But there are a variety of ways of measuring other properties of the atmosphere and the Earth's surface using sources of radiation on satellites. Until now these techniques have concentrated on radar methods using pulsed sources of microwave radiation, but in the future it may be possible to use lasers to do the same thing.

The three forms of radar equipment that have been extensively tested on experimental satellites are altimeters, scatterometers, and synthetic aperture radar. All

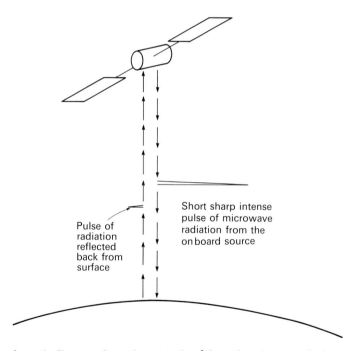

Pulse of radiation reflected back from surface

Short sharp intense pulse of microwave radiation from the on board source

Fig. 4.5. This schematic diagram shows the principle of the radar altimeter which measures the time taken for a pulse of microwave energy to travel from the satellite to the Earth's surface and back again.

work on the same basic radar principle which consists of sending out a pulse of microwave radiation and measuring both the time it takes for some part of that pulse to be reflected back to the satellite and also the intensity and frequency of the returned pulse. Each system uses different aspects of this basic principle to make different physical measurements.

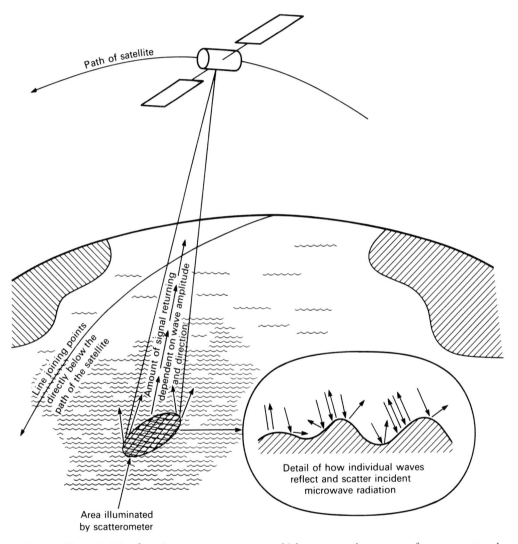

Fig. 4.6. The principle of a microwave scatterometer which measures the amount of energy scattered back to the satellite by the sea surface. The inset detail shows that the amount of energy reflected back depends on the size and steepness of the waves.

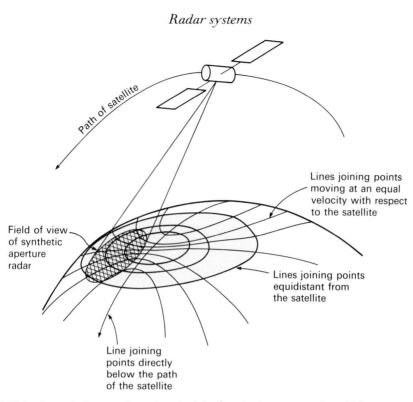

Field of view of synthetic aperture radar

Lines joining points moving at an equal velocity with respect to the satellite

Lines joining points equidistant from the satellite

Line joining points directly below the path of the satellite

Path of satellite

Fig. 4.7. This schematic diagram shows the principle of synthetic aperture radar which measures both the frequency shift and time taken for the signal to reach a point on the surface and return to the satellite. This enables the scattering properties of each element within the field of view to be uniquely determined.

The satellite altimeter is the most direct application of radar. It consists of a source of microwaves which produces a very short pulse of energy a few nanoseconds in duration, which is directed vertically downward to Earth. The time taken for the part of this signal reflected at the surface to return to the satellite (around 0.005 s) can be measured with sufficient accuracy to estimate the altitude of the satellite within a few centimetres. By knowing details of the satellite's orbit, it is possible to convert these observations into details of how the surface at any point beneath the orbit deviates from the geoid. Over the oceans this information can be converted into estimates of the strength of currents and other motions (see Chapter 10). In addition, roughness on the surface will blur the signal and the amount of blurring can be used to estimate wave height and from this infer wind speeds.

Scatterometers adopt a different approach. They spray out a relatively long duration pulse (a few milliseconds) in a fan-shaped beam ahead of the satellite to illuminate a narrow strip of the ocean. The pulse length means that the front edge of the pulse is most of the way back to the satellite before the trailing edge departs. The returning pulse is modified by scattering properties of the surface and the motion of the satellite which produces a frequency shift (the Doppler shift). By measuring both the amplitude and frequency profile of the returning pulse it is possible to make inferences about the direction and size of the waves on the oceans and from this estimate wind speed and direction (see Chapter 10).

A much more powerful technique is synthetic aperture radar (SAR). This combines the normal radar principle of measuring the time delay of the pulses returning radiation, with the frequency shift produced by the satellite's motion, to form images of the Earth's surface. The radar usually 'looks' to one side of the spacecraft in a direction perpendicular to its motion. It transmits a short pulse of microwave radiation toward the surface. Points equidistant from the radar are located on successive concentric spheres. The intersection of these spheres with the surface forms a series of concentric circles centred on the nadir point directly below the satellite. So, backscatter echoes from objects along a given circle will return to the satellite with an equal time delay. The points on the surface that produce equal frequency shifts to the return signal are all at the same angle to the direction of motion. This means they lie on coaxial cones, with the flight path as the axis and the radar at the apex. The intersection of these cones with the surface forms a family of hyperbolas. Objects on a specific hyperbola produce an identical frequency shift. The frequency shift is positive for points ahead of the radar and negative behind it. This means that within the field of view of the radar any point on the surface is uniquely defined by a combination of time delay and frequency shift. The amount of energy received from each point depends on the surface characteristics and so can provide an image of the surface. The resolution of the imaging system is dependent only on the measurement accuracy of the time delay and the frequency shift, but is independent of the height of the satellite. Typically, spatial resolution is about 25 m.

Because spatial resolution is so high, the formation of SAR images involves a huge amount of data processing and requires complex correction methods to handle the properties of the orbit and the effects of the atmosphere. The brightness of each pixel in the radar image is a direct measure of scattering properties of the area on the surface being viewed. This is mainly dependent on the roughness, the dielectric constant of the surface, and the frequency and the angle of incidence of the radar signal. Because surface parameters may vary considerably, radar characteristics have to be chosen to probe different surfaces.

5

Data handling

'As thick and numberless
As the gay motes that people the sunbeams'
John Milton 1608–74

HANDLING torrents of data, which are a feature of many aspects of meteorology, is a central problem for modern weather services. Indeed, the demands of numerical weather forecasting have been a major stimulus in the development of the latest supercomputers. Similarly, satellite meteorology has depended on advances in information technology. In almost every aspect of its development it has required extensive data handling, ranging from the collection and transmission of observations to the conversion of the raw data into useful input for computer models, and the preparation of visually informative images.

5.1 Collection and transmission

The starting point for the process of data handling in satellite meteorology is the continuous stream of signals flowing from detectors in the various on-board equipment. As seen in the preceding chapter, these detectors view solar and terrestrial radiation coming from a limited area below the satellite orbit. The amount of radiation is recorded at set intervals and converted into an electronic signal.

Each data point is encoded as a binary number. In the case of the AVHRR, for example, this is on scale from 0 to 2^{10} (1024) where zero-level corresponds to the deep space signal and full scale is set at the maximum anticipated signal for each channel. For the channels operating in the visible and near-infrared ranges this corresponds to a range of 0 to 100% albedo, where the signal is linearly related to the albedo. In the infrared channels, which measure terrestrial radiation, the equipment is designed on the assumption that the maximum signal will correspond to around 320 K. Because the observed signal is proportional to the fourth-power of the temperature of the Earth's surface or the atmosphere, the detector output normally has to be converted into equivalent temperatures before being used.

When the observations are used to form an image each data point can be used to form a single pixel. This defines the maximum spatial resolution that can be achieved. If, however, a higher signal-to-noise ratio is required than can be obtained from single observations, a number of adjacent points may be averaged to form a lower resolution but higher quality image. Alternatively, when requirements are less demanding the overall resolution and sensitivity may be reduced so that the imagery can be transmitted over low-cost data links.

Because of the high resolution and rapid scanning rate of modern radiometers each measurement interval is only a few tens of microseconds, and so, with multichannel instruments the observation rate can be as high as millions of bits per second. This is the basic information that has to be transmitted to Earth. Typically, these signals will use UHF (around 400 MHz) or S-Band (1600–2100 MHz) radio frequencies.

The signals obtained from viewing the Earth are only part of the data stream transmitted by the satellite. The instruments also have to be calibrated between each scan when not viewing the Earth. With the AVHRR, for instance, this involves detector readings obtained from viewing the on-board black body and space, together with four on-board measurements of the temperature of the black body. In addition, there are various timing and identification signals which are inserted at the beginning and end of the scan to enable the data to be processed automatically when it is received on the ground. This process uses identification signals to accurately define the position of successive scans in an individual image, and also allows the observations from different detectors to be interleaved in a single data stream, so that a number of images can be transmitted simultaneously.

Transmission of observations made by on-board instruments is just part of the overall data-handling and telemetry functions of weather satellites. In addition, they are used to collect data from automatic weather stations on land and in buoys, ships, and aircraft, and transmit these to a central station. Also, geostationary satellites are used for communications purposes between meteorological services. Finally, there is a complex telemetry system to provide ground control of the operations of the satellite to maintain its altitude and to adjust and control its instruments. The best way to consider these various systems is to describe those used on the current generation of weather satellites.

On the orbiting Sun-synchronous TIROS-N/NOAA series the output from the AVHRR is transmitted in real time and also stored by on-board recorders for transmission on command when passing over two receiving stations at Gilmore Creek, Alaska, and Wallops Island, Virginia. In addition, a station at Lannion, France, provides a back-up service. The real-time transmission of data consists of both high spatial-resolution images (1 km) and lower resolution images (4 km). These two sets of images enable different users to make immediate use of the satellite data. Major organisations, such as national weather services, will choose high-quality images, while smaller users, such as schools, colleges, universities, and a considerable number of private individuals with home-made receivers will use the lower resolution product. The on-board recorders, which collect data at the rate of over a million bits a second, store a complete global view of the low resolution imagery and a selected 10 min of the high resolution data. In addition they collect and transmit on command signals recorded by detectors in the atmospheric sounding instruments (TOVS).

Data-handling systems on geostationary satellites are broadly the same as for orbiting spacecraft, but with additional features. On Meteosat, in addition to two radio channels to transmit the data, there are 66 channels for transmitting data from automatic weather stations and eight channels for the exchange of processed data between users. As an indication of the growing scale of these operations the GOES satellites now have the capacity to handle data from up to 12 000 platforms per hour. These include not only standard automatic weather observing equipment, but river gauges, seismometers, and tidal gauges, all of which can either be programmed to transmit data at set times or interrogated by the satellite. In addition, where they are part of an emergency warning system, the satellites will automatically transmit alarm signals if observations exceed a prescribed threshold.

5.2   Assimilation and processing

The massive amounts of data transmitted by satellites must be subjected to a highly organised and thorough routine if they are to be properly used. Only with such a system is it possible to capture all the information and convert it into a form that is of both immediate use and also available for subsequent analysis. As a general principle, signals received at national control centres are passed immediately through a computer which strips out the telemetry signals and converts the data into a form that can be stored on computer-compatible files. In so doing, it calibrates the data and inserts Earth location parameters. Once this essential housekeeping and archival work has been done the raw data can either be passed directly to users or processed to provide standard products. All products require computer processing and frequently involve additional subjective analysis by scientists.

The data are processed in a range of ways at the various data collection centres to produce a huge variety of products. The nerve centre for US weather satellite data handling is the NOAA establishment at Suitland, Maryland. It processes data from both orbiting and GOES satellites. Atmospheric sounding measurements are converted into temperature profiles with figures for 40 levels, from ground level to an altitude of 65 km. In addition precipitable water in three pressure layers (up to 700 mb, 700–500 mb, and above 500 mb) is calculated. Both sets of figures are calculated globally on a grid-spacing of 250 km. Using the two current NOAA satellites this produces 16 000 separate sets of temperature and water vapour soundings each day which are disseminated almost immediately to 200 users around the world. The images from GOES-East are automatically analysed to produce low-level wind vectors on a 250 km grid three times a day, to provide 1200 observations daily.

Similar operations are conducted by other national weather services which collect comparable data from both orbiting satellites and other geostationary satellites or take the processed data from the US. All of this information, together with ground-based observations and those obtained from automatic platforms that are interrogated by satellites, is combined to form input to the standard weather forecasting exercises around the world.

The data which are not used in weather forecasting are produced at a more leisurely pace. Sea surface temperatures are generally available on a 25 km scale for local analysis within 8 h and an 8 km resolution of US coastal waters is available for the commercial fishing industry. On a wider scale global values of sea surface temperatures are available on a 100 km grid scale within 8 h. For climatological purposes, these are integrated to give monthly maps on a 250 km grid.

National data centres are only a small part of the information processing effort. High resolution images (1 km) from the NOAA satellites are collected by 80 receiving stations in 50 countries and lower resolution images (4 km) are picked up by 1000 receiving stations in 123 countries. In addition the visible and infrared images obtained from geostationary satellites by the major centres are, after processing calibration, re-transmitted through the same satellites, together with overlaid latitude and longitude grids and continental outlines, to secondary users. This process involves some loss of both spatial resolution and radiometric sensitivity. The visible images have a 5 km spatial resolution and 64 levels of brightness linearly related to the albedo of the surfaces in the field of view. The infrared images have an 8 km spatial resolution and have 64 levels of brightness nonlinearly related to the effective radiating temperature of the surface below. These images are recorded by several hundred users in the field of view of each geostationary satellite.

The pictures transmitted from either the orbiting satellites or re-transmitted via the geostationary satellites are used in a multitude of applications. Many of these will be reviewed in subsequent chapters, and all that needs to be said here is that some scientific groups have developed special systems which produce images that are more advanced than the national services. There are two groups that have developed this specialist skill to a high degree and whose work is featured extensively in subsequent chapters. The first is the Electrical Engineering Laboratory at the University of Dundee, Scotland. This laboratory has for over a decade produced the highest quality images of the weather over Europe and the North Atlantic from the NOAA satellite. The second group is the Space Sciences and Engineering Centre at the University of Wisconsin at Madison. Since 1970 the Centre has developed

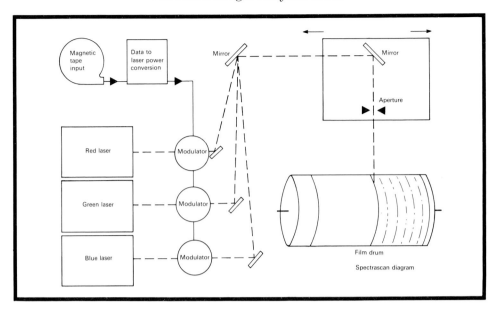

Fig. 5.1. False-colour satellite images can be generated from data on a computer using a set of lasers to translate the information on to a photograph.

the Man Computer Interactive Data Access System (McIDAS) system which combines orbiting and geostationary satellite data with conventional weather observations. This system can produce a wide variety of products which present the available data in a striking and accessible form. Its success can be gauged by the fact that McIDAS systems now operate at a variety of sites in the US as well as in Australia, Canada, Spain, and West Germany. But the visual appeal of such images is only part of the story. Their production also plays an important part in making the data accessible to both specialists and the man in the street.

5.3   Enhanced images and false colour

The role of high-quality images in advancing satellite meteorology is a classic case of improving data handling to present large quantities of information in a form that is more easily interpreted. The first stages of this process were to improve the presentation of the basic data so that weather features could be easily related to underlying surface geographical features. The inclusion of lines of longitude and latitude, and outlines of the land masses is now an automatic feature of data processing. When combined with the increased sensitivity of detectors, which have made it much easier to discriminate between land and sea, these techniques represented a major advance. Without such information many of the published pictures made only a limited impact on the viewer. But this type of enhancement is only the first step in a process that can be used to extract a great deal more information from the data.

Because most of the observations made from weather satellites are not in the visible part of the spectrum, it is open to the observer to present them in whatever form makes it easiest to visualise the physical processes at work. The simplest and most frequently used method is to employ a grey scale which runs from white for the lowest temperatures to black for the highest. Even greater visual impact is achieved by colour coding. Computer enhancement of images highlights features that are of particular interest to the viewer. This involves not only producing crisp, high-contrast black and white images or colour pictures which make the observed physical differences easy to pick out, but also using the computer to home in on certain physical aspects of the observations. This can be done, for example, by concentrating on a narrow range of temperatures to isolate developments at a certain level in the atmosphere. By using the entire dynamic range of the image formation process to represent differences at this level it is possible to gain a much more rapid insight of what is going on. Conversely, the use of colour may enable a number of different measurements (eg., of atmospheric temperature and water vapour) to be combined in a way which gives a better impression of the overall processes at work.

The demand for this type of high-quality colour picture has led to advanced techniques of transferring computer-stored data on to film. It is now possible to do this in one step by using the image data to control the power output of three lasers – blue, green, and red. This automatic system, which uses primary satellite data drawn from a computer-compatible archive to write directly on to film, achieves much higher relative and absolute adjustments of geometry and colour density. As such it effectively mirrors the scanning process by which the satellite-borne radiometers originally recorded the data.

5.4 Moving pictures

Another way in which computers can be used to present more visual information is to string a series of pictures from a geostationary satellite together to show the atmosphere in motion. Using infrared images it is possible to provide round-the-clock coverage of weather developments. This is of value not only to weather forecasters but also the general public, as such moving pictures shown on television can make the forecasts much easier to understand. This is particularly true of the latest development which uses temperature data to produce moving three-dimensional images. It also serves a scientific purpose as it can be used to provide an extended presentation of lengthy periods of weather observations. These movies have been used to study the way in which the weather patterns of the tropical Pacific varied during the dramatic changes in sea surface temperature that occurred in 1982–83 (see Chapter 10). These particular climatic fluctuations involved complex ocean–

atmosphere interactions over a period of more than a year. By using the entire sequence of geostationary pictures as a moving film it was possible to provide a clearer picture of how the patterns of cloudiness changed throughout the period, and to interpret how these changes were an integral part of variations in surface conditions.

5.5   Putting the data to good use

In their different ways, all the forms of data handling described here are integral to satellite meteorology. For much of the time this effort goes unnoticed, but it is important to recognise that the final products of the work bear the imprint of data processing. This imposes limits on what can be extracted from data in terms of sensitivity, selectivity, and resolution. So, in the examples that follow, it is important to keep in mind the underlying principles of the data presentation methods.

6

Interpreting satellite images

*'You don't need a weatherman
to know which way the wind blows'*
Bob Dylan

INITIALLY the most common published satellite images were of reflected sunlight. But, increasingly, infrared images obtained both by day and night are being used. We now see both types of image on television and in newspapers and magazines. They contain an immense amount of detail, most of which is too complex for the layman to interpret. But to meteorologists they can provide a lot of useful information at a glance. In this chapter we will consider just what can be extracted from these standard pictures and how any of us can get more out of them if we know what to look for.

From the moment the weather could be viewed from space, meteorologists' perceptions of it changed. The earliest photographs taken from rockets showed that the weather was much more complex than meteorological textbooks or standard weather maps could ever portray. The first satellite pictures amply confirmed how much detail about the weather could be obtained, from the local to the global level. These revelations produced what must seem contradictory responses. First, in emphasising the complexity of the weather they provided a major stimulus to make more detailed and thorough analyses of the atmospheric processes at work. Secondly, in spite of the detail, trained observers were able to confirm essential features in an image with which they could compare their analysis and check surface measurements.

Here we will explore both aspects of satellite imagery. The aim is to separate the essential active features of any weather situation from the intriguing but basically irrelevant detail of the broad picture. This discrimination is difficult to make, because what can be dismissed as clutter is physically real and can reveal a great deal about the processes at work. But for the purposes of defining what is going on and, more importantly, how the situation is likely to develop in the next 12–24 h, it has to be put on one side.

6.1   Standard images

Before considering how to interpret individual images it is as well to define a few of the basic features of the most frequently published pictures. The standard visible and infrared images from both orbiting and geostationary satellites look much alike, but do have a number of subtle differences. In all of them the most striking features are the clouds. They appear as white or light grey. Variations in the greyness do, however, reveal considerable differences in the state of the atmosphere.

In the visible range the brightest clouds are the deepest ones, which can scatter the most sunlight. The thinner the clouds the fainter the signal they produce, so it is not easy to tell the difference between thin low cloud and thin high cloud. In the mid-infrared the coldest areas are shown in the lightest shades and the warmest areas in the darkest shades. So the brightest clouds tend to be the coldest almost irrespective of their thickness, although the low emissivity of thin clouds is another factor which must be considered. This makes it hard to discriminate between high thin cloud and much thicker clouds. Moreover, in extreme cases low thin cloud, which may be close to the temperature of the Earth's surface, may not be seen at all.

Where the images show the surface the contrast between the two wavelength regions may be greater. In the visible, land shows up as lighter than the surrounding seas. But in the 10–12 μm infrared region, it is the temperature that matters. In summer, when the land is warm compared to the seas, it shows up as dark against a light background. Conversely, in winter the cold land can be lighter than the relatively warm seas which then appear as the darkest areas on the picture. But for the most part, it is the clouds that really matter when trying to make out what the weather is doing.

In coming to grips with the information in satellite images it is necessary to appreciate one central feature. Even at a relatively low spatial resolution of no better than 1 km, standard pictures contain a huge amount of data. In general, most of this cannot be used, and so sorting out the wheat from the chaff is essential. One way of understanding this problem is to consider a situation where the weather is in a quiescent state, but where there is a considerable amount of detailed structure in the distribution of clouds. A good example of such a weather situation occurred over the British Isles on 10 February 1980. The country was covered by a weak ridge of high pressure. The forecast was for dry conditions with sunny periods and only a chance of little rain in northwestern Scotland. This was correct but the country was covered by a complex orthogonal pattern of lines of shallow clouds. These clouds were triggered by the uplifting effect of both the day-time warming of the land and of the high ground acting on the fresh westerly winds. This generated not only lines of clouds perpendicular to the wind direction but also lengthy 'cloud streets' parallel to the wind that extended across the country. These demonstrate the fascinating atmospheric motions that can exist even on an apparently featureless day, and also show the problems facing forecasters trying to provide predictions of local conditions.

While this detail provides a fascinating insight in to the complexity of the weather, it is not what anyone using satellite images to make forecasts normally looks for. Instead, they tend to concentrate on major features associated with the main weather systems. These 'bare bones' will be examined in the examples described later in this chapter.

(a)

Fig. 6.1. Two examples of infrared satellite images of the British Isles showing the contrast between (a) *winter (27 February 1986) when the cold (white) lands shows against a warm (black) sea and* (b) *summer (16 May 1980) when the warm land shows dark against the colder sea (with permission of University of Dundee).*

6.2   Forecasting guidelines

Before considering the panoply of images of weather systems, there is benefit in investigating how they can be linked with what we experience on the ground. Because our perceptions of the weather are localised it is often hard to relate what is seen from space to our earthbound experience. The most obvious features are the brightness and degree of organisation of the major cloud structures. The brighter and more extensive they are, the more significant the weather events they represent.

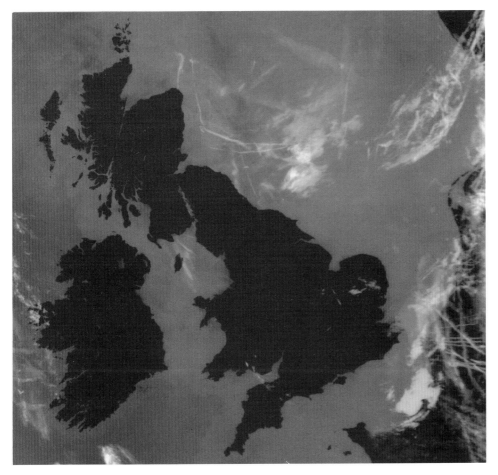

Fig. 6.1. (b)

Having picked these features out, the next stage is to relate them to what is being presented by the forecaster. This is being made ever easier by the presentation of sequences of geostationary satellite images in the form of moving pictures, together with predictive sequences of what may happen during the next 12–24 h. Nevertheless, interpretation can be improved by remembering the following basic guidelines:

(1) at the most elementary level the well-defined features of mature low pressure systems will tend to continue their current motion in a relatively orderly manner, undergoing little significant change over periods of 12–24 h. So where, for example, a cold front is crossing the country in a given direction it is reasonable for the observer on the ground to expect to see the succession of weather types in the satellite image passing overhead. This is the basic

Fig. 6.2. An image of the British Isles on 10 February 1980 showing that, even when there is no significant weather about, complex cloud patterns can develop (with permission of University of Dundee).

76

process that is used to produce predictive sequences using a succession of geostationary images and extrapolating the cloud movement up to 12 h ahead;

(2)   where low pressure systems are still in the early stages of growth, the standard models (see Chapter 2) provide a useful guide of how the combination of fronts will change over a day or two;

(3)   where the weather is undergoing more subtle changes (eg., rapidly developing small depressions, thunderstorms brewing up, or fronts dissipating) the basic rule is to concentrate on what the forecast says about developments and apply it to the relevant features on the satellite images.

These basic observations only scratch the surface of the information that can be extracted from published satellite images. Weather forecasters and trained observers have a whole series of guidelines to help interpret current weather and they use satellite images to add depth to the forecasts generated by computer. But for the layman, concentrating on the most basic aspects of interpreting pictures may help in understanding the wealth of material that is spread before us daily, and may also assist us in extracting more from the examples of the weather systems that follow.

6.3   Depressions – large and small

Low pressure systems with characteristic closed circulation come in all sizes from quickly developing small wave depressions to huge extratropical and tropical storms. While many of their features are different, as described in Chapter 2, their visible images bear many similarities. Here we will examine the links between these dynamic features which show up so clearly on satellite pictures.

Where better to start than with the most extreme example ever recorded from space – typhoon Tip which developed in the western Pacific in early October 1979. By 12 October it had developed into a tight circular pattern and a clearly defined eye. The central pressure had by then fallen to 870 mb – the lowest value of sea-level pressure on record. The temperature of the cloud tops around the central eye was an extraordinarily low −93 °C, a clear measure of the great height to which the clouds extended. At this stage the forecasting information available in the satellite images did not focus on the individual features of the storm, but its symmetry. With its well-defined eye and high degree of concentricity it was clearly a particularly vigorous storm. When information on cloud-top temperatures was added it became evident that this was an intense cyclone which would cause immense damage if it made landfall. As it happened Tip vented most of its energy over the ocean. Over the next six days the typhoon matured as the central pressure rose and the storm extended over a greater area. The view from space showed a majestic system with an area of organised circulating intense convection extending over 2000 km.

Vigorous extratropical depressions exhibit the same sense of symmetry and massive scale of circulation. These properties can clearly be seen in an image of what was possibly the deepest low pressure system ever observed in temperate latitudes. This developed off the coast of Nova Scotia, Canada, on 13 December 1986. By the morning of 15 December it filled most of the Atlantic north of 50° N latitude. Its centre was sited between Greenland and Iceland and had an atmospheric pressure of less than 920 mb. The satellite image at this time shows a huge well-developed spiral. Because the storm was so large the fronts extend off the image and the most

Fig. 6.3. The most intense typhoon (Tip on 12 October 1979) ever measured in the Pacific. This infrared image shows the extraordinary symmetry of intense tropical storms. At the time the central pressure in the clearly defined 'eye' of the storm was 870 mb (with permission of Japanese Meteorological Agency).

Fig. 6.4. In its later stages, Typhoon Tip (18 October 1979) spread out to form a giant depression as this visible image shows (with permission of Japanese Meteorological Agency).

striking feature is the effect of the huge plunge of arctic air behind the storm. This intensely cold air sweeping southward over relatively warm water triggered off widespread convection and masses of shower clouds.

Over land, depressions lose some of their grandeur, but if anything the features of the standard model are easier to see. A well-developed low over the Great Lakes on 4 March 1985 provides a good example. In the image the warm front has advanced over New England and the cold front stretches from Michigan to Texas. The most important difference from sea-centred depressions is that behind the cold front there is no cluster of heavy showers, as land does not have the heat capacity to trigger off convection in the same way as over the oceans. This explains why over the continents the weather is so predictable after the passage of a cold front associated with a vigorous low – there is almost inevitably an extensive clear slot following along behind. Whereas, when a front is crossing the ocean or moving onto land, as is often the case in northwestern Europe, forecasting the timing and intensity of any showers after the initial clearance is more difficult.

This brief summary of different types of depressions can be completed by considering those small features that are part of a developing train of major lows. These secondary features start as no more than a thickening of the clouds on a cold front trailing back from the original low pressure area. But they may grow rapidly so that within 24 h they can become fully fledged systems in their own right, as the example cited here from 11 March 1982 demonstrates. The decaying initial area of low pressure is centred over Iceland. The secondary consists of the rapidly expanding area of cloud to the west of Ireland, which is already showing clear signs

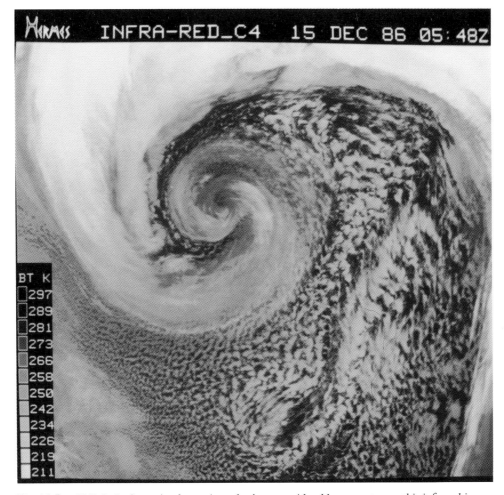

Fig. 6.5. In middle latitudes major depressions also have considerable symmetry, as this infrared image of the North Atlantic (15 December 1986) shows. At the time the central pressure of the depression was below 920 mb. The important features are the spiralling cloud system in the central region of the storm and the associated plunge of arctic air which generates a region of vigorous showers (with permission of UK Meteorological Office.)

of developing a well-organised circulation. In the next 24 h this system behaved as forecast. It deepened rapidly and moved swiftly across Scotland and into the North Sea. In so doing it brought gales and heavy rain to Scotland and colder air down across the British Isles.

These small systems not only bring some of the most dramatic weather developments, but they can grow into the dominant systems in global circulation patterns. As such they are an integral part in the process by which an endless succession of atmospheric eddies transports energy to high latitudes in both hemispheres (see Section 2.2.2). Accordingly, they can be considered as no different than other low pressure systems. All are major actors on the meteorological stage, but in terms of what can be seen from space they are only part of the plot.

6.4 Fronts, thunderstorms, and tornadoes

It would be an oversimplification to focus all the attention on the major features of well-defined depressions or rapidly growing new low pressure systems. Much of the most dramatic weather we experience develops out of what may seem to be insignificant features in limited areas of the larger weather systems. To understand how these events can materialise out of seemingly clear air we must first extend the discussion of the characteristics of the fronts associated with mobile depressions. We have seen that the major elements of low pressure systems, notably cold fronts, centres of circulation, and convective activity behind cold fronts are often well defined. Starting with these apparently predictable aspects of depressions, the motion of cold fronts often proves easiest to interpret, because their intensity is largely governed by the uplifting effect of the push of the wedge of cold air from behind. Consequently, they are frequently narrow, being clearly marked by a cloud mass 100–200 km wide and perhaps 1000 km long, and moving in a relatively uniform way over the course of 24 h or so. In particular, over land this means that their passage often tallies well with local forecasts and can be matched up with what is seen in published satellite pictures. Not only is their timing predictable but so is the succession of weather, with increasing cloud and precipitation followed by a sudden clearance. The latter can lead to startling falls in temperature, with 'cold waves' notably in the US or China. Here, arctic air can sweep south and east unimpeded and without warming up appreciably so that changes in temperature of up to 30 °C in a few hours can occur.

But, sometimes they do not develop according to plan. Apparently vigorous fronts can fade away, or, weak features may pick up steam suddenly, especially over the oceans. Modern forecasting is becoming increasingly effective at foreseeing such changes. None the less, where satellite pictures have been obtained some hours after the forecast has been completed, closer inspection of this new information can be used to refine weather predictions. Tell-tale signs such as a cold front showing evidence of thinning out and becoming braided foretell declining vigour. Conversely, particularly bright thick fronts provide support for predictions of heavy precipitation. Moreover, local thickening can presage a secondary system which not only produces more widespread weather, but also delays or even reverses the apparently predictable forward motion of a front.

Such simple observations cannot, however, be made about warm fronts, or more generally about the warm sector of low pressure systems. This is because active warm fronts are difficult to locate on satellite cloud pictures. The general region of the front can often be seen in a well-organised cloud band but, within the large area the clouds cover, the position of the front is difficult to pin-point with any accuracy. Decks of clouds at different levels disguise the actual line of contact between air masses of different origins and physical characteristics, although as a broad guide the surface front is normally found some 700–800 km behind the leading edge of the cloud band.

Another problem with interpreting cloud images is that they tell us little about the stability of the atmosphere. The most favourable regions for severe local storms involve troughs of low pressure in the lower and middle troposphere, copious moisture in the lower troposphere, and converging winds near the surface. Regions of convergence are often associated with former atmospheric boundaries such as warm fronts and the remnants of earlier thunderstorms, fresh boundaries created by new storms, or interactions between storms and the broader atmospheric flow

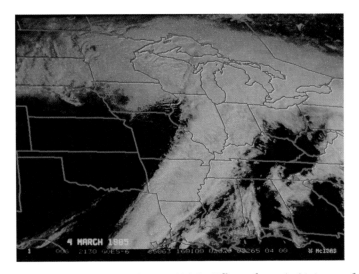

Fig. 6.6. Over the continents depressions take on a slightly different form. As this image of a major storm (4 March 1985) over the Great Lakes shows, the fronts are more clearly defined as there is no significant convective activity behind the cold front stretching from Michigan to Texas (with permission of University of Wisconsin).

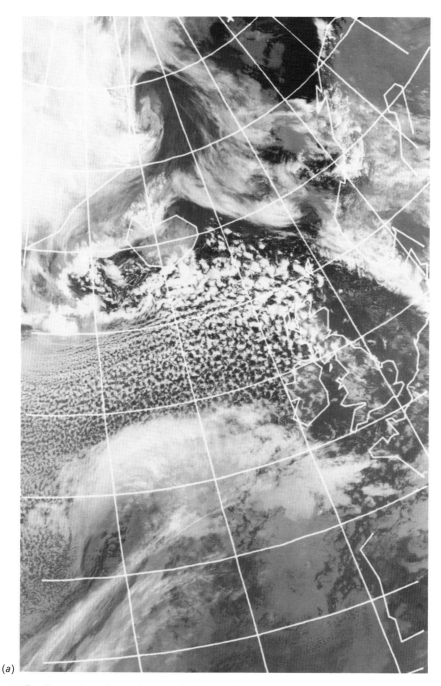

(a)

Fig. 6.7. Small secondary depressions can develop rapidly to become major systems as this pair of images show. The developing low in region of 25° W 50° N at 0934 GMT on 11 March 1982 (image a) deepened rapidly as it moved across the British Isles in the next 24 hours to be in the northern North Sea by 0909 GMT on 12 March 1982 (image b) (with permission of University of Dundee).

83

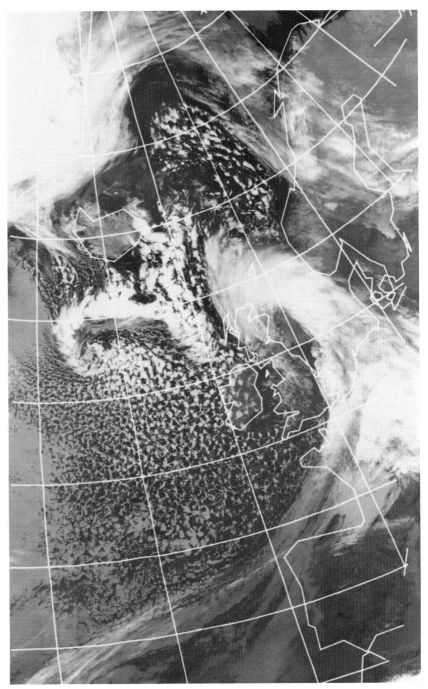

Fig. 6.7 (b)

pattern. To these features can be added the effects of unequal heating of the land surface by solar radiation. Cloudy areas will stay cooler during the day than places where the sky is clear. In summer months this means that clear areas can warm rapidly. This uneven heating generates vigorous convection which can lead to the growth of storms in a matter of hours over areas that were cloudless in the morning.

The ways in which the stability of the atmosphere can be detected from space will be discussed in the next chapter. For the moment the instructive features to look for in published images are bright discrete blocks or blobs of cloud standing out clearly against a dark background. They can be isolated single features or more organised strings of storms advancing on a broad front. A good example of the latter

(a)

Fig. 6.8. The rapid development of thunderstorms can be seen in this pair of images. Taken at (a) 1255 GMT and (b) 1435 GMT on 14 July 1982 they show the growth of storms on a cold front over northern France and the Netherlands (with permission of University of Dundee).

occurred with a thundery low west of France over the Bay of Biscay in mid-July 1982. As it moved slowly northerly its cold front had produced up to 100 mm of rain and severe flooding in southwest England on 12 and 13 July. By late morning on 14 July the cold front stretching back across Belgium and down to the Alps was no more than a few high clouds. In the next 2–3 h a string of intense thunderstorms erupted all along the apparently quiescent front. This shows just how quickly severe weather can brew up where the atmosphere is unstable.

Even more dramatic storms can develop over the mid-western US. The clash of humid air from the Gulf of Mexico, and dry cold continental air often produces some of the most intense thunderstorms in the world. These storms can develop in an organised way to produce a clearly recognisable structure. Here a line of convective clouds associated with some weak wider-scale atmospheric feature can pro-

Fig. 6.8 (b)

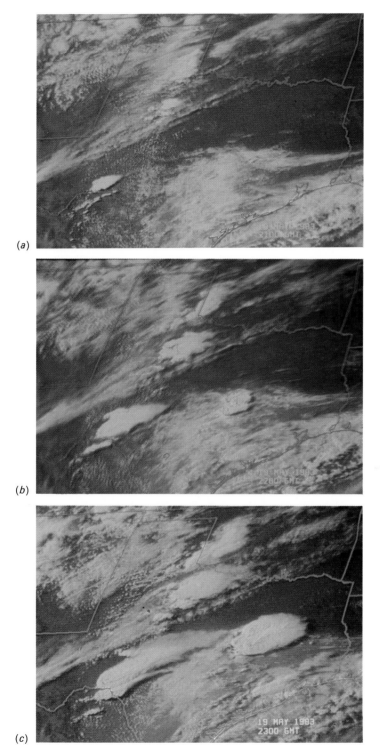

Fig. 6.9. The growth of severe storms which can produce tornadoes can be seen in these three images. Taken (a) at 2100 GMT, (b) 2200 GMT and (c) 2300 GMT on 19 May 1983 over Texas they show how rapidly these storms can develop (with permission of University of Wisconsin).

duce a front of its own. As the rain falling from the parent thunderstorm cools the air, a cold downdraft occurs. This cold blast flows outward along the ground and, if the air is unstable ahead of the cold air boundary, a cloud line appears and thunderstorms develop along the line. These new storms, in turn, add more cold air to the advancing cold air boundary and the cloud line continues to propagate and generate new clouds and thunderstorms.

This type of recognisable organised convective activity is used by forecasters to predict the development of severe weather a few hours ahead. It is of particular value in advising civil aviation authorities of the potential danger to aircraft of windshift, or strong low-level winds and wind shear being encountered. Each year aircraft are lost in such conditions, such as the tragic case of the crash of a Tristar in August 1985 at Dallas, Texas, with the loss of 133 lives.

An even more frightening feature of these storms is that some of them spawn tornadoes, which cause tragic loss of life and damage to property in many parts of the United States. Satellite pictures offer a way of improving tornado warning systems. In recent years, certain cloud shapes have been shown to be reasonable indicators of tornadoes. In particular a bright V-shaped cloud wedge with overshooting towers of convection is a signature of such devastating storms. These are recognisable by the bright and dark wave-like areas which occur where the tops of exceptionally vigorous thunderstorms protrude through the cirrus anvil that normally caps intense convective activity, are illuminated by the sun, and cast shadows on the cirrus below. Good examples of such images were recorded on 19 May 1983 when violent tornadoes caused widespread destruction and loss of life in eastern Texas and around Houston.

6.5 Global patterns

Thus far, in interpreting standard satellite images, we have focussed on aspects of individual weather systems. This does, however, neglect the global picture, and it is in providing a complete view of the Earth's weather patterns that satellite meteorology may have most to offer to the layman. Whether it is in the early mosaics which showed the weather patterns as if seen from above the poles, or the geostationary pictures from above the equator, these pictures make it much easier to see how the atmosphere is forever shifting energy toward the poles.

Polar projections prepared from mosaics of images collected from orbiting satellites confirm observations that have been made about the general behaviour of extra-tropical depressions. At any time, a view from the North Pole will show a series of eddies whirling northward with fronts trailing behind, so that the whole process looks like a vast atmospheric Catherine-wheel. At first there seems to be little rhyme or reason to the pattern of these eddies, but on closer inspection a certain order can be deduced.

To appreciate these patterns we must recall what was said in Chapter 2 about the mid-latitude jet stream and the long waves in the upper atmosphere. These waves are an essential part of the process of poleward energy transfer. They can be visualised in a polar projection of weather patterns by remembering two facts: first, the clouds usually denote areas of rising air and the clear regions are where air is descending; and second, the centres of circulation are usually pushing eastward and northward. So the jet stream tends to skirt round to the top of mid-latitude depressions and dip southward in the clear areas. This reinforces the process of drawing warm air toward the poles and feeding compensating bouts of cold air toward the equator.

Another feature that stands out clearly on the polar projections is the change from winter to summer in the Northern Hemisphere. The most striking contrast is the presence or absence of continental snow cover. In winter much of North America and northern Asia has persistent snow cover. Another difference is that the strength of the mid-latitude westerly circulation weakens appreciably in summer. So in winter the depressions are more numerous and vigorous.

At lower latitudes both polar projections and pictures from geostationary satellites tell the same story. The largely clear skies over subtropical latitudes mark the descending arm of the Hadley Cell and also pick out many of the major deserts of the world. In equatorial regions of the intertropical convergence zone there appears an irregular necklace of areas of intense convection girdling the globe. Depending on the season it will move a few degrees north or south following the overhead Sun, but whatever the time of year it is always clearly seen. In addition, in summer more extensive and organised cloud formations can be seen extending northward over the Indian subcontinent – clear evidence of the monsoon. One other striking feature of the geostationary pictures, especially over the Pacific, is that at almost any time of the year in one hemisphere or the other, there will be a tropical storm.

6.6 Going into more detail

These preliminary observations about the content and interpretation of satellite images merely skim the surface of the subject. Not only can experienced meteorologists extract basic satellite imagery but also, when other measurements are made, the combination of observations can lead to interesting complications. In considering the increasingly sophisticated instrumentation and data analysis involved, it is important not to lose sight of the basic features of the weather revealed in these standard images. Many of the more advanced sensing techniques, notably those which measure the temperature of the atmosphere and the Earth's surface, have to trade spatial resolution for accuracy – a single observation may cover an area several tens of kilometres across. This inevitably leads to a loss of detail and with it some of the glorious array of forms of atmospheric behaviour. This loss of detail is further compounded in computer forecasting work where the resolution is typically no better than every 100–200 km (see Chapter 12). So at all times it must be remembered that many important features occur at a scale of only a few kilometres across, a scale which is far beyond the capability of the most advanced computer models of the global climate. From now on we will be considering more manageable quantities of information, but at the cost of working with models that can only show the basic features of the global weather machine.

7

The structure of the atmosphere

'Wild air, world mothering air,
Nestling me everywhere.'
Gerard Manley Hopkins 1844–89

THE FUNDAMENTAL PROBLEM with satellite images considered in the preceding chapter is that for the most part they are obtained by measuring reflected sunlight. As such they tell us relatively little about the vertical structure of the atmosphere. While both a knowledge of the three-dimensional nature of weather systems and the use of infrared images permits us to infer quite a lot about what is going on at various levels in the atmosphere, only so much can be extracted from such basic images. The various sounding systems described in earlier chapters enable us to measure the temperature, moisture, and motion of the atmosphere at different levels, and thus gain much more insight into the processes at work. To understand just what can be observed it is easiest to concentrate on specific examples measuring particular weather conditions.

7.1 Images of terrestrial radiation

In the previous chapter standard infrared images were discussed briefly. These produce pictures based on measurement of the level of terrestrial radiation which are in many ways comparable to those obtained by measuring reflected sunlight. While this form of imagery has the advantage of presenting many of the features of the weather in a recognisable form it does not use all the information available in the radiometer observations. Because these contain comprehensive details of the temperature of the cloud tops they can be converted into false colour images which give a more complete picture of the weather systems. Moreover, because the colour coding can be chosen to cover any specific temperature range, it is possible to produce images which highlight certain slices of the atmosphere.

At first glance false colour images may appear confusing, and hence they are rarely used in published weather forecasts. But to the trained observer they offer an immediate and far greater insight into the structure of current weather systems compared with standard models. This allows meteorologists to add detail to their forecasts and also, in combination with other measurements, to refine their models. A good example of this process of analysis and development is in the case of hurricanes.

Early satellite images of hurricanes showed that their intensity was linked to their symmetry. This led to a set of forecast rules about the degree of organisation of tropical storms and what this meant in terms of their likely development. Initially, these relied on simple visual and infrared images. However, infrared temperature measurements of cloud-top temperatures in the form of false colour images provided additional information about symmetry and hence degree of organisation of the storm. They also added completely new data on cloud height, which is a direct measure of the intensity of convection in the core of the hurricane. This information provided forecasters with an extra dimension in their analysis of the potential impact of wind damage and rainfall – as the amount of damage is proportional to the square of the velocity of the wind speed. Since damage due to individual hurricanes in the United States can run to more than $2 billion and an average of two hurricanes per year make landfall, any improvement in forecasting is of potentially great economic benefit.

False colour images can also reveal information about extratropical depressions, by highlighting certain slices of the atmosphere. The standard model of these depressions described in Chapter 2 focussed on the development over the lifetime of the system. Equally important is the dynamic pattern of the depression at any given time, as the strength of the various features is a guide to the weather that can be expected to be associated with their passage.

Of particular importance in a maturing depression is the rising current of warm air pushing up ahead from the warm sector of the low. The strength of this conveyor belt of moisture in the middle troposphere is an indicator of the amount of precipitation associated with the development. As noted in Chapter 6, visible images have difficulty in picking out the essential features of the warm sector of the depression and the position of the warm front. However, infrared images, which highlight certain levels in the atmosphere, allow forecasters to form a clearer impression of the structure of cloud bands and the amount of moisture they contain and so refine predictions of local weather.

This improvement in short-range forecasting of rainfall associated with strong convective activity is examined in Chapter 12. But here it is important to realise that cloud-top temperatures obtained from infrared images produced by geostationary satellites provide excellent means of detecting the development of fast-growing shower clouds and thunderstorms. As these progress during the day, the half-hourly pictures show their growth so that they can be detected before they become significant weather features.

7.2   Temperature sounding

Chapter 4 identified snags involved in using the measurement of the upwelling radiation from the clear atmosphere to derive a vertical temperature profile. These are both theoretical and practical. At the theoretical level a first guess must be made of what the temperature profile is before the calculation can be refined. The further this guess is from the truth the greater is the possibility that a seriously flawed solution will be obtained. At the practical level the greatest impediment is clouds, but even when these can be corrected for, or solved using microwave radiometers, the complexity of the atmosphere causes difficulties. In particular satellite observations cannot detect sudden changes in temperature with height such as marked inversions due either to intense radiation cooling at the surface or mixing of air of different origins at higher levels. This can lead to significant errors. Furthermore, if the tropopause is abnormally high or low, similar problems can occur, because the first guess will assume it is at the normal level for the time of year.

Because the magnitude of any error is critically dependent on the specific atmospheric conditions it can vary greatly from day to day. When the profile is close to the climatic normal and shows a monotonic decline throughout the tropopause agreement between satellite sounding and radiosonde measurements is close to the criterion of ± 1 K sought by meteorologists for computer forecasting work. At the other extreme, errors of 10-times this size can occur. The most comprehensive set of tests conducted on the accuracy of satellite temperature sounding methods was during the First Global GARP Experiment (FGGE), run throughout 1979. The results of this work are given in Table 7.1.

On average the errors are a factor of two or three greater than the target of ± 1 K specified for the GARP (see Table 3.1). The greatest errors occur at the surface and near the tropopause. Thus the problems associated with low-level temperature inversions and the location of the tropopause are critical factors in improving the performance of sounding techniques. The decline in performance in cloudy conditions confirms both the restriction of the infrared techniques and also the limited sensitivity of microwave radiometers.

The general conclusion from these observations is that infrared soundings are better in clear conditions while microwaves are the only way to probe cloud-covered areas. Improvements have, however, been made in recent years by combining the results of the two sets of observations. This is particularly important in eliminating poor data, as in forecasting work inaccurate observations are worse than no observations at all. The combination of the two techniques has proved of considerable value in areas of well-broken cloud where microwave results can be used to check the quality of the infrared observations. This is, however, only part of the problem, as the errors are not random. They are greatest in cloudy areas, which is just the place forecasters most want to know what is happening. So the real challenge is to explore more accurately the temperature structure inside active weather systems. The only way that this will be achieved is to improve the performance of microwave sounding methods. Microwave radiometers are now being designed to probe the atmosphere in more detail with the objective of providing this additional information.

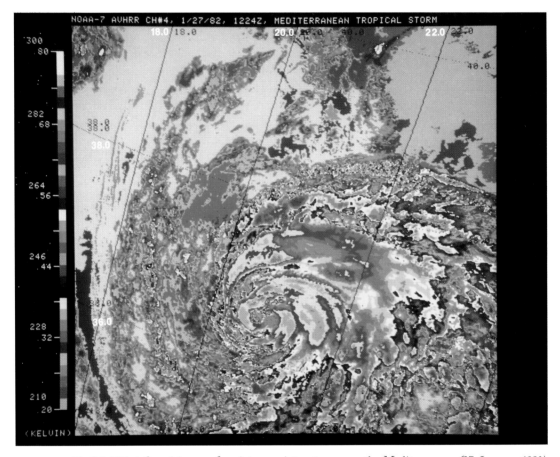

Fig. 7.1. This infrared image of an intense winter storm over the Mediterranean (27 January 1983) shows how the use of false colour can be used to highlight the temperature structure of the system − reproduced from Weather *with permission of the Royal Meteorological Society.*

94

Table 7.1. *The route mean square difference (in K) between satellite retrievals and radio-sonde data observed during FGGE in 1979*

Atmospheric layer (mb)	Atmospheric conditions		
	Clear	Partly cloudy	Overcast/cloudy
100–70	2.2	2.2	2.2
200–100	2.1	2.2	2.3
300–200	2.3	2.5	2.8
400–300	2.3	2.4	3.0
500–400	2.3	2.4	3.0
700–500	2.0	2.1	2.7
850–700	2.4	2.7	3.4
1000–850	2.8	3.1	3.8

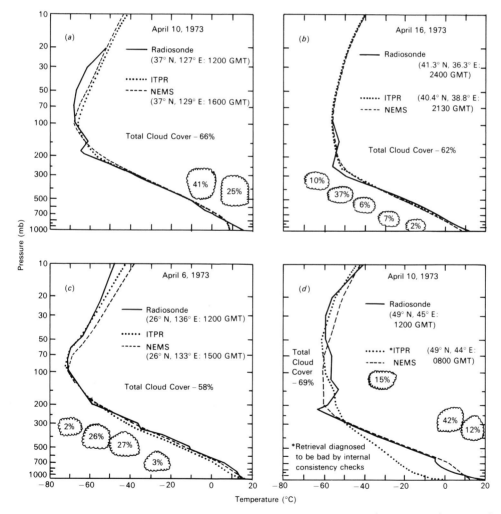

Fig. 7.2. *This comparison between Nimbus 5 temperature soundings derived from NEMS (microwave) and ITPR (infrared) observations with radio-sonde observations shows that under certain conditions satellites are capable of making accurate measurements of atmospheric temperature profiles (after Smith* et al. *(1974) in Houghton* et al. *(1986).* Remote Sounding of Atmospheres. *Cambridge: Cambridge University Press).*

7.3 Ground-level temperatures

Satellite data cannot produce information about the detailed structure of the lowest few hundred metres of the atmosphere. But, since we spend our lives here, such measurements are of central interest to us. The problem is that the air close to the surface, often termed the 'boundary layer', has different characteristics to the rest of the atmosphere. Because of the radiative properties of the surface and its effective roughness the temperature and wind speed in this layer may be radically different to that of the air just a few tens or hundreds of metres above.

In spite of these problems, and the fact that errors in temperature measurement are greatest at the surface, physically important observations can be made. Using the most transparent portions of the 'window' regions infrared radiometers can make widespread observations of maximum and minimum temperatures which are comparable with the standard ground-level measurements made from a standard Stevenson's shelter, but where standard thermometer readings are available it is not normal practice to use satellite observations for this purpose. There are, however, occasions when satellite observations can provide considerable insight into what is happening at low levels. For example, when it is exceptionally cold and the radiative contribution of the atmosphere is minimal, satellite observations can provide considerably more information than ground-based instruments.

A good example of the ability of satellites to measure ground-level temperatures was during the cold spells of December 1981 and January 1982 in Britain when new records for minima were established ($-27.2\,°C$ at Braemar, Scotland, on 10 January and $-26.0\,°C$ near Shrewsbury, in the West Midlands, also on 10 January). The synoptic situation during the first cold spell was a basically northerly flow over the country, while the second spell was easterly with an anticyclone centred over Scandinavia for much of the time. The extreme temperatures were recorded under clear skies 2–5 days after extensive snowfalls. The remarkable feature was that in the same locality (the Shrewsbury area) both spells featured temperatures of $-25\,°C$ and $-26\,°C$ which broke long-standing records for England and might statistically have been expected only once in 100 years.

Images formed by combining observations made by the AVHRR on NOAA 7 using the three 'window' channels at 3.7, 11, and 12 μm show temperatures over Britain in the early morning on 10 January 1982. The lowest temperatures are readily seen as the whitest areas on the image and coincide with the valleys. In fact, the level of brightness is a reflection of the topography, except over Lancashire and Yorkshire which were the only snow-free areas. What this shows is that the coldest conditions occur where cold air can drain into lowlying areas. Even more interesting is that surface temperature measurements were in almost exact agreement with screen minima measured on the ground. This, despite the fact that standard climatological data are taken at a height of 1.2 m and minima usually occur 2–3 h later than when the satellite image was recorded. The net effect appears to be that the differences cancel each other out.

These observations tell us a considerable amount about what produces exceptionally low night-time temperatures in Britain and in other parts of the world. The essential atmospheric ingredients are clear skies, very dry air throughout the troposphere, and an almost flat calm at the surface. These conditions maximise radiation loss from the surface, minimise downward radiation from the atmosphere, and ensure that a stable temperature inversion is formed near the surface. In addition the presence of freshly fallen snow of moderate depth greatly enhances radiation losses and minimises heat conduction from the soil, while large catchment areas allow the coldest air to flow downhill and collect by the process known as 'katabatic drainage'.

The value of these observations extends well beyond simply explaining exceptional conditions during an isolated cold spell in Britain. They show that the principal reason for the record-breaking cold was the extreme dryness of the air. This explains why such low temperatures are more common over the interiors of northern continents and shows the moderating effect of the oceans on temperatures in the British Isles. It also confirms the importance of the insulating and radiative properties of snow – in areas of the UK where there is no snow cover screen temperatures rarely fall below $-10\,°C$. Satellite observations could thus be used in providing warnings to frost-sensitive industries such as agriculture and horticulture.

This type of warning is already in regular use in Florida. Between 1977 and 1985 the eastern US experienced a string of exceptionally hard winters with frequent cold waves sweeping down from the Canadian Arctic. In particular, four exceptional cold waves hit Florida between 1980 and 1985. The effect of these frigid outbursts is clearly reflected in the change in acreage of citrus groves. During the first half of this century citrus groves were expanded to cover a peak of 950 000 acres in 1970. Since then, however, they have contracted to around 760 000 acres. While property developments are partially responsible for this decline, the weather has played its part. Damaging cold waves have, on average, affected Florida every four or five years during the last 100 years or so, but were less frequent between 1920 and 1960. The more recent frequent severe spells seem to have no equal since the Spanish explorers first introduced orange trees 400 years ago.

These changes in the incidence of frosts in Florida not only raise important questions about long-term variations in the climate but also place a greater premium on accurate forecasting. Some protection can be provided by hosing the trees with water, running diesel heaters, or operating electric wind machines. All of these methods are extremely expensive, costing some $5 million per night to the industry. So, they can only be justified economically if there is a high risk of the temperature falling below the damaging threshold of 28 °F ($-2.2\,°C$). Given that in cold winters there may be as many as 60 nights when temperatures fall below freezing in northern Florida, potential savings are large. Using GOES infrared images NOAA have been able to make observations of temperature to an accuracy of $\pm 1\ °C$ every half-hour, and from this provide timely warning to the orange growers. This has led to more efficient use of frost protection systems. However, when the temperature falls as low as $-13\ °C$ in citrus-growing areas, as it did in January 1985, there is no way to save the trees from irreparable damage.

7.4   Urban heat islands

Another use of satellite surface-temperature measurements has been to confirm the well-known effects of urbanisation. Ground-based studies have shown that major urban areas have striking impacts on local meteorology. The greatest changes occur during calm clear nights when temperatures in surrounding rural areas may fall by as much as $10\ °C$ more than at the centre of the city. Using high resolution infrared satellite imagery it is possible to produce heat maps. The most illuminating results have been obtained with the Heat Capacity Mapping Mission on the NASA Applications Explorer Satellite which was launched in April 1978 and operated until September 1980. It was designed to conduct a series of environmental measurements and carried a two-channel radiometer which operated in the 0.5–1.1 μm region and in the 10.5–12.5 μm region. Because it had a spatial resolution of 0.5 km it could produce detailed maps. These show clearly that the effect of urbanisation on night-

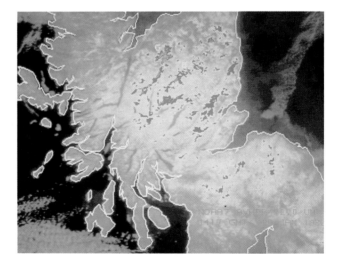

Fig. 7.3. Satellite radiometers are capable of making accurate measurements of surface temperatures. This infrared image of northern England and Scotland on 10 January 1982 shows that the lowest temperatures occurred in the lowest-lying areas (coloured magenta) and that the values recorded by the radiometer agree closely with standard surface measurements (with permission of UK Meteorological Office).

time minima is greatest in the largest conurbations, but it is also evident that even quite small towns produce night-time warming significant enough to be seen from space.

7.5 Fog and air pollution

While both low cloud and fog interfere with observations of the temperature of the lower atmosphere, they can easily be detected in visible images. Observations of how such conditions form, are sustained, and then break up can be of considerable meteorological interest. Fog has a characteristic appearance of flat texture, abrupt edges and, over land, patches and patterns conforming to terrain features such as valleys, and water boundaries of rivers, lakes, and reservoirs. Radiation fog, which is a common occurrence during autumn and winter nights in mid- and high latitudes, in particular behaves in this manner. It can be widespread, notably when it forms under clear night sky conditions which allow maximum loss of radiation. Generally, radiation fog occurs as a relatively thin layer and dissipates during the morning hours.

Analysis of satellite visible imagery has shown that brightness variation in fog areas is a measure of the thickness of the fog. Also, the images show that fog dissipates from the outer edges toward the denser central areas, with the thickest parts clearing

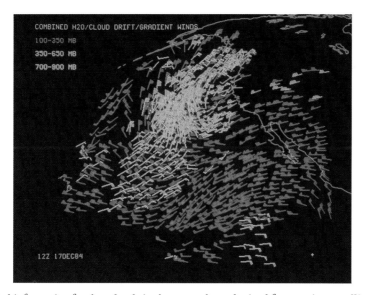

Fig. 7.4. Wind information for three levels in the atmosphere obtained from various satellite sources can be combined in a single image to provide a better understanding of the weather. This image shows data for 17 December 1984 over the eastern Pacific Ocean (with permission of University of Wisconsin).

last. This behaviour may seem obvious, as the effect of solar heating in the morning will be to generate vertical mixing that eats away at the edges of the fog bank. But, until observed by satellite, it had never been confirmed physically. This enabled meteorologists to improve their predictions on how fog will clear during the day.

The formation of air pollution often follows the reverse pattern. During the summer months, most notably over the eastern US and to a lesser extent over industrial regions of Europe, stagnant weather can occur which can result in poor air quality and reduced surface visibility. Air pollution and haze are recognisable in visible imagery by a grey appearance which makes ground features less distinct. So the build-up of pollution can be monitored during the day. Early in the morning, during stagnant episodes, the pollution may be no more than a few streaks in cloud-free areas. By mid-afternoon it can extend and thicken dramatically. Even more interesting is that the haze bands may act as boundaries along which thunderstorm activity can develop. Afternoon convection often forms along the boundary of a dense haze area, where there is discontinuity of surface heating, as it is cooler under the haze and warmer in the clear areas. Not only does this provide an operational forecaster with a guide as to where thunderstorms can form, but it also raises interesting questions about the effect of air pollution on the weather. There have been indications for some years that increases in atmospheric pollutants have led to increases in rainfall. It may be that future satellite observations will provide the evidence needed to resolve this question.

7.6 Winds aloft

So far there is no way of obtaining direct information about the speed and direction of wind as a function of altitude from satellite observations. Such information is central to providing an accurate picture of the current state of the atmosphere. At high latitudes this gap can be largely filled by detailed measurements of the pressure and temperature of the atmosphere which can be combined with the physical laws of motion to produce a reasonable representation of weather systems. But in the tropics where the Coriolis force due to the Earth's rotation is negligible the motion of the atmosphere cannot be reliably calculated. So knowledge of the winds is essential to improved forecasting in the tropics.

Wind information can be obtained by tracking cloud displacements using successive images from geostationary satellites. But this approach has one fundamental limitation – it cannot make measurements where there are no clouds, and even where there are clouds the levels being tracked may not move with the general wind. Differences of up to $\pm 40\%$ occur between wind velocities which would be deduced by tracking a single cumulus cloud in its growing stage or its dissipating stage. These effects are greater at high levels but inevitably introduce errors of between ± 1.5 and $\pm 2\,\mathrm{m\,s^{-1}}$.

High clouds also mask low clouds and so it is difficult to get measurements at different levels over the same spot. In addition, the emissivity of high thin clouds is not known. This means that the apparent temperature of the clouds does not correspond to that of the adjacent atmosphere. So it is difficult to calculate cloud height with any accuracy. This is frustrating as the clouds provide useful information of winds at high levels ahead of warm fronts and associated with thunderstorms. These winds control the movement of such features and so are good indicators of their future movement.

In spite of these limitations winds associated with weather systems at low levels ($\simeq 900$–850 mb) and at high levels ($\simeq 300$–200 mb) are obtained as a matter of routine with an accuracy of about ± 6 m s^{-1} and can in some circumstances be considerably better. But this falls a factor of three short of what weather forecasters would like for their work. Moreover, research has shown that with more careful classification of cloud types it is possible to obtain additional information about winds in the mid-troposphere. But to produce measurements over a wider area and at different levels more sophisticated equipment is needed.

Two particular approaches to obtaining better estimates of wind speeds at different levels are currently the subject of active investigation. First the water vapour channel on Meteosat, which observes upwelling radiation in the 5.7–7.1 μm region, has been successfully used to greatly extend the coverage of wind measurements. Successive images of this radiation can be used to track the movement of regions of different relative humidity in the atmosphere. These clouds of water vapour extend more widely than the places where condensation occurs and so reflect movements over wider areas. Moreover, this technique can more accurately define the level of the water vapour and so is used to provide wind fields at three levels in the troposphere. It has proven particularly valuable in mapping the course of the jet stream. But the accuracy of wind speed measurement is no greater than that obtained from measuring cloud movements. The second technique uses the VAS instrument on the GOES satellites. Because this can make simultaneous measurements of both the temperature profile of the atmosphere as well as the position and temperature of cloud tops it is possible to obtain more accurate measurements of motion at different levels in the atmosphere.

While many limitations of wind speed measurements remain, the greater confidence of defining the clouds as being either low, medium, or high is a significant improvement. Furthermore, any information on wind speed in tropical areas where it is not possible to make inferences from measurements of temperature and surface pressure is a great advance. But, none of these developments can address the question of the wind speed close to the Earth's surface. Over land we will have, for the foreseeable future, to rely upon ground-based observations using standard anemometers. Over the oceans satellite microwave techniques, which are discussed in Chapter 10, offer far greater promise.

8

Measuring rainfall

'When it rains it rains on all alike'
Hindu proverb

FOR CENTURIES scientists and meteorologists have kept records of rainfall. The earliest measurements were probably made in Korea during the fifteenth century. Since the late seventeenth century measurements have been made in Britain, and a series of monthly rainfall figures for England and Wales has been constructed from data collected since 1727. However, many of the early observations were unreliable because of the sites chosen for rain gauges, and recent surface measurements, though more accurate, could only sample a tiny part of any rainfall pattern. So, while the collection of more and more statistics ironed out many of the fluctuations and gave a reliable climatological picture, it could not uncover much of the detail of the precipitation process.

The essence of the problem of rainfall measurement is the rapid change in precipitation over both space and time. Apocryphal stories of teeming tropical rainfall which soaks one side of the street and leaves the other dry, are only an extreme version of the fundamental difficulty facing meteorologists. At the other extreme, in mid-latitudes, rainfall is generally very light and only a low percentage of events involve heavy rain. By the same token, the social implications of these various forms of rainfall are equally varied. A little light rain can disrupt sporting events but be of no hydrological significance. Warnings of intense bursts of rainfall can be of immense importance in taking preventive action against damaging floods, while in agriculture reliable rainfall forecasts over the growing season is important.

These requirements have, however, been only part of the demands made on meteorologists. All the significant records of rainfall have been made on land. Virtually nothing is known about the amount of rain that falls over the seas. While such measurements may seem of little importance to our lives, they are in fact of considerable meteorological and climatic importance. They provide a direct measure of the energy released in the form of latent heat in weather systems over the oceans. This is of interest in studying individual systems, and is also an important parameter in understanding energy release in the tropics which drives the Hadley cell and acts as the tropical 'boiler' running the global atmospheric heat engine.

The vast range of requirements concerning rainfall, both in terms of measurement and forecasting, impose conflicting pressures on weather services. Satellites offer the opportunity to tackle these problems in a number of original ways. In particular, geostationary satellites provide observations of all precipitation systems in their field of view every half-hour or so. While these can provide a great deal of information about the extent and depth of cloud systems, they are a far cry from making direct measurements of rainfall. To do this meteorologists must either form reliable relationships between cloud types, as identified from satellite, and rainfall rates or develop direct ways of measuring rain as it falls.

8.1   Cloud brightness and temperatures

Some of the first pictures from TIROS were analysed to relate cloud images and ground-based radar measurements of precipitation. A study of the pressure systems over the mid-western US on 20 April 1960 showed that correlation between cloud cover and rainfall patterns was not straightforward. But as data accumulated over subsequent months it became clear that the brightness of the clouds might be an important factor. This is not surprising, as on the basis of a simple physical analysis the amount of light scattered by a deck of cloud is proportional in some way to its thickness. However, this relationship is complicated by the shape of the clouds and the angle of the Sun.

Throughout the 1960s a great deal of effort was devoted to relating images obtained from space with surface classification of clouds, to identify those most likely to produce rainfall. This in itself was not an easy task, but it was made even more difficult by the growing realisation that the stage of development of the cloud systems played a crucial role in the amount of rainfall. The brightest clouds appeared to produce more rain in the early stages of their development than when they had reached full maturity. This meant that orbiting satellites could only provide limited information. Regular observations from geostationary satellites were needed to keep track of cloud development. Furthermore, ground studies confirmed that along with this variation over time, the rainfall was usually confined to only a small part of the area covered by the clouds.

Parallel difficulties were associated with attempts to use infrared measurements of the temperatures of the cloud tops to predict rainfall. While ground measurements showed that the coldest temperatures were associated with the most widespread precipitation, in general the results were disappointing. In part this was because high cold clouds did not ensure the presence of deep clouds. Better results were obtained when visible and infrared data were combined to detect cold bright clouds. Over the years it was possible to build up an index of cloud types and refine the interpretation of the images to obtain increased accuracy. In 1975 comprehensive tests in the United States concluded that using GOES satellite images, 45% of the estimates of rainfall fell within 1 mm of surface measurements, 72% within 3 mm, and 95% within 13 mm. While these techniques represented the major rain areas fairly well, the overall effect was to spread out the rainfall patterns. So, low rainfall was often overestimated while heavy rainfall was underestimated. Also daily spatial variability was smoothed out, but the results improved with integration over time and space.

The other main prong of the attack on the problem of satellite rainfall measurement has been microwave measurements. But before considering progress in this area it is useful to consider the parallel development of ground-based remote sensing techniques using radar. Because much more work has been done on such surface measurements, it provides a good introduction to the problem of improving observations from space. Moreover, much of the calibration of satellite observations has used ground-based radar measurements.

8.2 Radar measurement of rainfall

The discovery during World War II that rainfall scattered microwave radiation and hence produced measurable radar signals led to a whole new area of meteorology. In the 1950s and 1960s this subject grew rapidly. It built on both theoretical analysis and practical experience of correlating radar observations with surface measurements of rainfall rates. Both approaches illustrated the complexity of the processes involved.

Theoretical analysis shows that the scattering properties are dependent on both the number of droplets in a unit volume and their size distribution. The amount of radiation scattered also depends on its frequency as the refractive index and the absorption coefficient of water varies with frequency. In theory this should enable accurate predictions to be made of the scattered signal. But in practice the size distribution is a complicated function of the nature of the rainfall. In particular it underestimates rainfall due to drizzle, which has no large droplets. In addition, the refractive index and absorption of ice are radically different to those of liquid water, so melting snow produces dramatically higher signals.

In addition there are practical problems. Because radar systems cannot make accurate measurements close to the ground they fail to see shallow precipitation. Conversely, they fail to see evaporation below the beam when rain is falling through a dry layer of air near the surface. Radar also overlooks effects which occur when air rises over hills. This produces low cloud which is missed by the beam but which

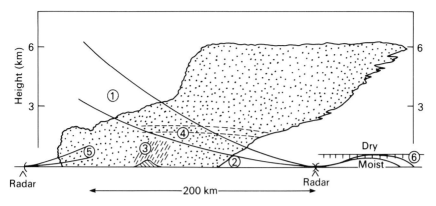

Fig. 8.1. *The measurement of rainfall using surface radar is complicated by a number of factors. As shown on the diagram these include: (1) radar beam overshooting shallow precipitation at long ranges; (2) low-level evaporation beneath the radar beam; (3) orographic enhancement above hills which goes undetected beneath the radar beam; (4) anomalously high radar signal from melting snow; (5) under-estimation of the intensity of drizzle because of the absence of large droplets; and (6) radar beam bent in the presence of a strong hydrolapse causing it to intercept land and sea.*

leads to a process of orographic enhancement as it greatly increases the amount of rain falling.

The best results obtained by radar suggest an accuracy of $\pm 20\%$ for hourly totals in a catchment area of 100 km². While this may not seem particularly impressive it is a considerable advance on scattered surface measurements. But these can only be obtained within a radius of 50 km of the radar transmitter and in cases where melting snow is excluded. None the less, these results provide a benchmark, not only for surface work, but also of what might eventually be possible from space using radar techniques.

8.3   Microwave measurements

In principle the best approach to measuring rainfall from satellites would be to use radar techniques, but this involves high power levels and so early attempts to measure rainfall from space using microwave techniques used a different approach. Because raindrops have complex absorption and scattering properties at microwave frequencies they will also have distinctive emission properties at these frequencies. So a microwave radiometer viewing the emissions from clouds will detect different signals depending on whether or not the cloud contains an appreciable number of raindrops.

The first instrument designed to examine the microwave emissions of clouds was launched by the Russians on COSMOS 243 in 1968. While this instrument only worked for two weeks it was able to construct latitudinal profiles of the liquid water content of the atmosphere. The first American orbiting microwave instruments went aloft on Nimbus 5 in 1972. In particular, the Electrically Scanning Microwave Radiometer (ESMR 5), provided the first measurements of rainfall from space. This instrument made measurements at a frequency of 19.35 GHz. This frequency was chosen so as to avoid the effects of water vapour and oxygen in the atmosphere and concentrate on the emission from liquid water droplets, but it was not capable of discriminating rain over land, because of the varying emissivity of the land. None the less, the radiometer was capable of detecting the high emissivity (high radiant temperatures) of rainfall over the oceans and made interesting measurements of the development of precipitation associated with tropical storms.

The next stage of development was the ESMR 6 launched on Nimbus 6. This worked at 37 GHz, which, although twice as sensitive to water vapour and oxygen in the atmosphere, was three times as sensitive to liquid water. This enabled it to make measurements over land. In addition, it was able to discriminate the polarisation of the upwelling radiation. This had the advantage of being able to detect surface wetting and flooding which otherwise was confused with heavy rainfall.

The more sophisticated five-channel scanning microwave radiometers flown on Nimbus 7 and SEASAT have advanced these techniques still further. Observations made in the North Atlantic as part of the SEASAT programme showed that liquid water and rain-rate distributions inferred from space were in qualitative agreement with surface measurements. But there is still a long way to go before reliable passive microwave measurements of rainfall can be made from space.

So far the only attempt to replicate successful ground-based radar techniques of rainfall measurements on satellites has been the active scatterometer system flown on SEASAT. This instrument demonstrated that it was possible to detect rainfall, but during the short lifetime of SEASAT it was not possible to obtain reliable quantitative comparisons with surface measurements.

8.4   Practical results

In describing the efforts to develop reliable methods for measuring rainfall from space particular emphasis has been placed on the difficulties that have been experienced. But this does not mean that useful results have not been obtained. The measurements may have limitations but where there is nothing better they are of huge potential benefit. This is particularly true over the less developed and sparsely populated areas of the world, and over the oceans the results represent a completely new source of meteorological information.

In places as far apart as Oman, Sumatra, and Surinam interpretation of satellite images has led to improved hydrological data. Satellite observations are capable of providing better estimates of rainfall which can be used to assess planned projects, and to assist in their subsequent operation. For instance, in Surinam satellite data have been used to improve the operation of a hydroelectric scheme and associated flood control system. In general these results show a good match with the limited surface data, but are inclined to smooth the peaks and troughs in the rainfall amounts.

Satellites have done much to improve understanding of precipitation processes over the oceans. In the tropics they have produced new insights into how convective activity seems to demonstrate a degree of order not anticipated from ground observations. In particular, the association between areas of convective and heavy precipitation, and changes in surface temperature in the equatorial Pacific, associated with the El Nino (see Chapter 10) have been studied. During the period 1979–81 microwave observations of liquid water content of the atmosphere showed a good correlation with climatological data. The estimated accuracy of values obtained for seasonal rainfall was $\pm 30\%$ during this period. With the El Nino in 1982–83 microwave measurements showed a drastic reduction in rainfall over the oceans around Indonesia, in the tropics of the South Pacific, and on the northern edge of the ITCZ, and a substantial increase over the normally dry central and eastern equatorial Pacific.

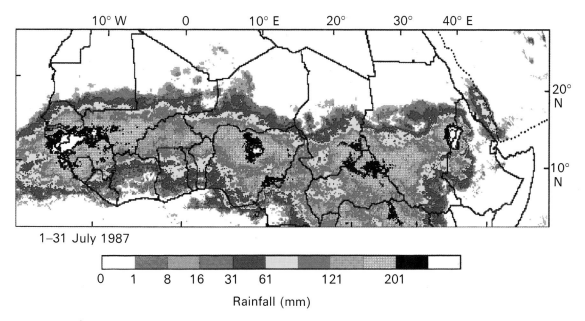

1–31 July 1987

0   1   8   16   31   61   121   201

Rainfall (mm)

Fig. 8.2. Rainfall estimate prepared from statistics of cold clouds for July 1987 using Meteosat infrared data by TAMSAT Group at the University of Reading (with permission of Department of Meteorology, University of Reading).

Satellites have detected previously unknown rainfall patterns over the southern oceans. They have revealed that in the southern Atlantic in the region between 25 and 50° W longitude and 25–50° S latitude, rainfall is considerably higher than elsewhere in this latitude band, suggesting that this area is a source of the depressions that are the major climatic feature of the latitudinal band around 40° S. This behaviour may be linked to the influence of the Andes on the long-wave patterns in the upper atmosphere (see Section 2.2.2).

Interesting results have also been obtained in estimating the amount of rainfall associated with severe storms which cause flash floods. Two examples that have been examined closely are the storm that caused the flood in Big Thompson Canyon, Colorado, on 31 July 1976 killing 139 people and the storm in Johnstown, Pennsylvania, on 19–20 July 1977 which killed 77 people and did $200 million damage. In both cases, satellite imagery provided reasonable estimates of rainfall rates, although there was a tendency to smooth out the most extreme figures. These results suggest that future efforts to improve short-range forecasts to provide warnings of such extreme weather events (see Chapter 12) will rely heavily on the use of satellite observations.

GOES images have also been studied to see whether they can be used to produce better forecasts of heavy snowfall in the mid-western US. An examination of 75 storms occurring between 1979 and 1984 showed that it was possible to obtain clues about the intensity of snowfall from subtle features in the satellite pictures. When combined with surface data and the strength and movement of the storm, these results demonstrated that it should be possible to make improved predictions of heavy snow.

Although all of this may seem like slow progress the potential benefits are beginning to emerge. For instance, the TAMSAT Group (Tropical Agricultural Meteorology using SATellite and other data) at the University of Reading, England, which is funded by the UK Overseas Development Administration, the Food and Agricultural Organisation (FAO), and the European Economic Community produced interesting results for the Sahel region by comparing satellite images with a network of surface instruments. These techniques have used statistical rules to estimate rainfall on the basis of the duration of cloud tops being below certain temperatures (eg., -40, -50, or -60 °C). While local results on any given day are not reliable, averaged over periods of 10 days or more and for areas of several thousand square kilometres, the data provide figures which are useful in agricultural planning. They have also shown that satellite data are better for making hydrological estimates of run-off in river basins to improve flood warnings, which is of particular value in sparsely populated arid areas where rare heavy storms can wreak havoc.

9

Measuring the surface of the land

'Ye are spies; to see the nakedness of the land ye are come.'
Exodus 42:9

IN MANY BASIC ANALYSES of the weather, changes in land surface receive relatively little attention. While it is accepted that the oceans play a central role in the climatological patterns of the Earth and that changes in the extent of snow and ice can play an equally important part, variations in features such as vegetation cover and soil moisture are given little attention. This is not entirely surprising as the atmospheric consequences of such changes are on a smaller scale. Nevertheless, they are important, and may often play a dominant part in local weather patterns and may even be crucial in certain wider forms of climatic change. For instance, it is estimated on the basis of satellite measurements that the average surface albedo of the British Isles rose to a value of 0.24 in the long hot dry summer of 1976 as compared with a figure of 0.14 in more typical moister years. This is a significant change which altered the radiative properties of the surface appreciably (see Section 2.1.4). Furthermore, in areas of meteorology such as agricultural advisory services which combine current weather with an integrated analysis of the effect of past events, the accuracy of measurement of the surface conditions is a vital part of sensible predictions.

9.1 Global vegetation index

At the most general level satellite measurements can provide a global inventory of the photosynthetic activity associated with growing vegetation. This can be achieved with the AVHRR on the NOAA satellites using the visible (0.55–0.68 μm) and near-infrared (0.725–1.1 μm) channels. Growing vegetation absorbs strongly in the visible region (0.4–0.7 μm) but scatters strongly in the near-infrared, so by comparing signals in the two channels an estimate can be made of the amount of vegetation present (see Section 4.3.6).

This measurement is complicated by a number of factors. Such effects as low Sun angle, off-axis radiometer observations (which have higher atmospheric absorption), and the presence of aerosols and clouds all serve to reduce the contrast between the two channels. Also at temperatures close to or below freezing, photosynthetic activity virtually ceases and this shows up as a reduction in absorptivity in the visible region. To ensure that the amount of active vegetation is not underestimated weekly observations are evaluated and the highest contrast figures are considered to be the best measure of the amount of photosynthetically-active vegetation. In addition, no measurements are made at temperatures below 0 °C.

In spite of these limitations comprehensive measurements have been made of the amount and seasonal change of vegetation in tropical and temperate latitudes. Not only have these provided a clear picture of seasonal and annual geographical changes in photosynthetic activity, but they also offer a means of monitoring the uptake of CO_2 from the atmosphere. In due course it may be possible to use the measurements to obtain a better understanding of the part vegetation plays in delaying the climatic effects of man-made carbon dioxide.

9.2   Desertification

The awful consequences of the drought in the Sahel in the early 1970s brought the impact of the southward spread of the Sahara to the attention of the world. At the UN Conference on Desertification held in Nairobi in 1977, it was estimated that deserts were spreading at a worldwide rate of $60\,000\,km^2\,y^{-1}$. By this time satellite measurements had already played a significant part in the debate about the causes of these changes. Images of the Israel–Egypt border across the Negev Desert, or of rangelands fenced off in Mali provided clear evidence that herds of browsing animals belonging to nomads played a crucial part in the removal of the last vestiges of vegetation. Clearly, satellite monitoring provided the best way of measuring the advances of the desert and also of estimating the physical effects of any such changes. This work remains critically important as, despite the calls for action in 1977, all the evidence suggests that desertification has continued unabated in the last decade.

Key to Fig. 9.1 (opposite)   Normalized difference vegetation index

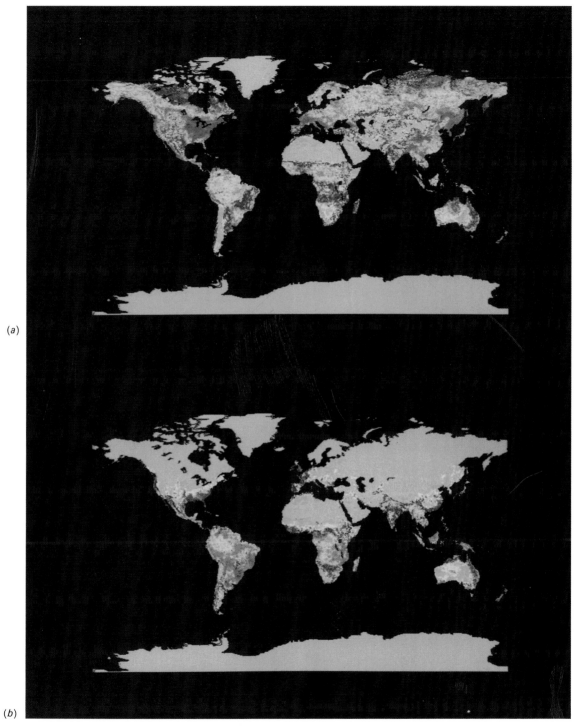

Fig. 9.1. The change in global vegetation cover can be measured using weather satellite observations. These two images show the difference between the extent of vegetation in (a) January and (b) July (with permission of NASA).

The climatic consequences of desertification have been highlighted in various efforts to model the effect of changes of albedo in the southern Sahara. The initial approach to this issue was to calculate the impact of increasing the albedo in the Sahel from a value of 0.14, typical of savannah, to 0.35 for desert sand. The effect of this change was to produce reduced summer rainfall in the Sahel region. This feedback mechanism, which reinforces the effects of drought, seems to be the product of less absorption of solar radiation at the surface and a consequent stabilisation of the atmosphere that reduces convection and hence rainfall. This theory has been supported by recent observations along the Mexico–US border. Measurements using GOES images have shown that clouds grow 50–177% more rapidly to the north of the border. This difference is attributed to US grazing practices which protect vegetation cover more effectively.

There is, however, greater doubt about whether overgrazing can lead to permanent climatic shifts. Analysis of Meteosat images during the slightly wetter period around 1979 showed that in certain areas such as Senegal and Upper Volta, vegetative cover increased, reducing reflectivity and hence counteracting climatic effects of earlier desertification. These observations contradict the assumption that once vegetation is removed the irreversible nature of the climatic processes ensures it will not regenerate. More sophisticated models are needed to analyse changes in both the albedo and the hydrologic cycle associated with desertification. Moreover, the subsequent severe drought in the early 1980s suggested that such processes were not only local events but were probably part of global changes, notably of sea surface temperatures. In particular, warmer conditions in the tropical South Atlantic may be critical, and the El Nino (see Section 10.3) in the tropical Pacific may also have played a part. So the climatic impact of desertification must be considered in both the regional and global context. Furthermore, the link between local effects and wider climatic events is particularly suited to satellite monitoring.

9.3 Dust storms

Although the amount of dust in the atmosphere is not strictly a surface phenomenon it is intimately linked with aridity and hence is most easily considered alongside the question of desertification. Satellite observations have proved particularly effective in measuring the presence of dust in the atmosphere. Such measurements are not normally part of standard meteorological procedures, so comprehensive data on the frequency of dust storms in arid parts of the world and the extent of the dust clouds emanating from these areas are not available.

In the United States both the NOAA and GOES satellites have been used to monitor the development and movement of dust storms in the plains states. For example in February 1977 a storm originating in New Mexico and Colorado was tracked to Alabama and Georgia and out into the Atlantic. This was the worst such storm since the dust bowl years of the 1930s. In the Middle East satellites have been used to observe storms that would otherwise go undetected. This is important because of their impact on oil-field operations, shipping, and military activities. The effect of the drought in the Sahel has been studied by measuring the movement of dust clouds across the tropical Atlantic.

9.4   Deforestation

The increasingly rapid removal of forests in the tropics in recent decades has been the subject of widespread concern to ecologists. For the most part this concern has focussed on the impact on the flora and fauna of the equatorial regions, and relatively little attention has been paid to the potential climatic consequences. It has, however, been suggested that the removal of the forests in West Africa could be yet another explanation for the drought in the Sahel. It is argued that cutting down the trees could have reduced the amount of evaporation in this region and hence cut the amount of rainfall. This is yet another area where satellite observations may hold the key to understanding climatic change.

A comprehensive survey by NASA and the Lamont–Doherty Geological Laboratory, New York, of both historic records and satellite images has been conducted to estimate anthropogenic changes in vegetation cover in West Africa during the last century. The survey concluded that over half the forests (about 1 million km²) have been removed and replaced by agriculture activities. Satellite measurements suggest that the climatic effects have probably been small because the new vegetation cover absorbs about the same amount of sunlight as the forest it replaced. So, despite large-scale deforestation in the region, the impact on both the local climate and, more important, the Sahel to the north appears to have been negligible.

Similar conclusions about the climatic neutrality of tropical forest clearance have been reached from other studies. Even where vegetation cover is stripped and the soil left bare the effects appear to be small, since the increase in the amount of sunlight reflected from the cleared areas is balanced by a reduction in the formation of clouds because less water is pumped into the atmosphere where there are no trees.

9.5   Measuring evaporation

As has already been discussed, changes in surface conditions affect the rate of evaporation. Not only does this have potentially significant consequences in terms of atmospheric feed-back mechanisms but it is also important for weather forecasting. The amount of evaporation from the surface will affect the temperature of the lower atmosphere and hence its stability. Knowledge of how much water vapour is being created at the surface is of value in predicting the intensity of convective activity and the formation of damaging thunderstorms.

Estimating evaporative rates is complicated by the heterogeneous nature of the Earth's surface. Croplands have varying roughness, albedo, root depth, and stomatal characteristics. These properties affect both the amount of water vapour produced and the surface wind speed. When moisture is plentiful most of the net absorbed radiation is converted into the latent heat of vaporisation, whereas when moisture is scarce it will be converted into sensible heat, warming the land and the adjacent atmosphere. Although the rise in surface temperature is affected by wind speed and atmospheric stability, as a first approximation it will provide a measure of the proportion of incident solar energy that is converted into sensible heat. So, if the albedo is also measured, the remaining energy can be assumed to have been converted into latent heat. This means, for example, that measurements made by the GOES satellites using visible images to estimate albedo, and by VAS to infer temper-

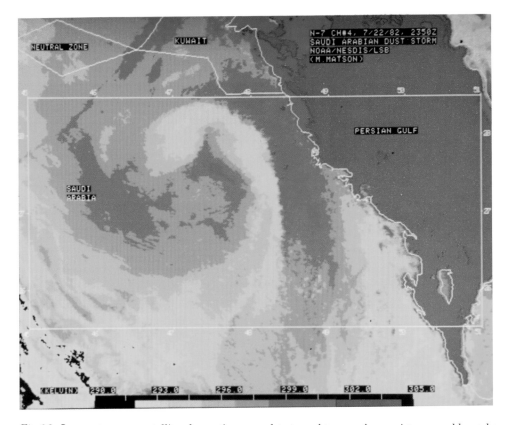

Fig. 9.2. In remote areas, satellite observations can detect sandstorms whose existence could not be inferred from sparse surface observations. This example of a sandstorm observed at nighttime in northern Saudi Arabia (22 July 1982) shows a comma-shaped feature associated with the storm (reproduced from Weather *with the permission of the Royal Meteorological Society).*

Fig. 9.3. The extent of tropical forest clearance can be measured using satellites, as these Landsat images of Rondonia Province, Brazil, in 1976 and 1981 show (with permission of NASA).

atures at the surface and through the atmosphere can be used to estimate the amount of evaporation.

Given a knowledge of the local rainfall in the area it should, then, be possible to calculate the water budget and hence soil moisture levels, assuming that during the summer, run-off and deep infiltration do not dominate the depletion processes and can usually be estimated following heavy rainfall. Such estimates of soil moisture are of considerable value in predicting crop yields and advising farmers on the most efficient irrigation strategies. Thus far only experimental work has been done on such techniques, but these methods can be expected to develop to supplement the operational services that already exist for agriculture in many parts of the world.

9.6 Agricultural forecasts

Weather is the most important and variable factor in year-to-year fluctuations in agricultural production. Forecasting ultimate levels of production depends on reliable weather observations. These provide estimates of soil moisture, crop yield, and crop stress. The accuracy of the forecasts is reduced when weather observations are sparse or are not received quickly. Satellites offer a solution providing they can produce sufficiently accurate measurements of precipitation, daily temperature extremes, canopy temperature, insulation, and snow cover. As discussed elsewhere, most of these variables can be monitored from space, but the way in which they combine to produce forecasts and how these are checked against observations of crop development is an additional refinement.

Weather satellites are not ideal for many aspects of crop monitoring. While the AVHRR on the TIROS-N/NOAA series has produced interesting global observations, the resolution limit of approximately 1 km is too coarse for accurate measurement of specific crops. Higher resolution can, however, be obtained using Earth resources satellites (eg., the Landsat series) which are not normally used for meteorological purposes, as they survey a narrow swathe ($\simeq 180$ km) and only return to the same location every 18 days. They can, however, make surface measurements with a resolution of better than 100 m. When combined with surface and satellite weather observations these results can be used to make improved forecasts of crop yields.

These forecasts have become increasingly influential in the behaviour of commodity markets – the Russian grain purchasers who bought up huge quantities of cheap grain on the Chicago market to cover the disastrous Soviet harvest of 1972 showed the value of having advanced warning of low yields. Similarly, Landsat pictures demonstrated the true extent of the killing frosts in the Brazilian coffee-growing areas in July 1976, which led to an almost doubling of coffee prices around the world. Subsequent successful US forecasts of annual yield during the string of poor harvests in the Soviet Union in the late 1970s and early 1980s provided added confirmation of the value of this work.

One more local agricultural aspect of monitoring vegetation cover is in developing locust plague control. The primary factor in the growth and movement of swarms in Africa and the Arabian peninsula are a combination of temperature and moisture which can lead to the rapid growth of vegetation. The UN Food and Agriculture Organisation have developed a scheme to use satellites to monitor vegetation so that pre-emptive spraying can be conducted. The first success based on this method was in southern Algeria in August 1981, and the use of satellite data is now an essential element in the fight against locusts.

9.7   Flooding

The economics of flood protection is a subtle balance between the probability of damaging events and the cost of preventing such damage. The worldwide cost of such events is huge – in the United States alone they frequently exceed $1 billion per year. In planning how to reduce the chance of flooding in an economical way engineers and planners need improved information of the areas at risk and the potential extent of inundation. Satellites provide an accurate and comprehensive means of assessing the overall extent of flooding and hence provide a better guide to its economic consequences.

Flooding shows up best on night-time thermal infrared imagery, when thermal contrasts between land and water are greatest. Satellite pictures can be used to identify extensive floods where large rivers burst their banks and inundate large areas of flat land. This information is of considerable use to rescue services. Research on these techniques was conducted during the 1973 Mississippi River floods and the 1978 Kentucky River and Red River floods in North Dakota and Minnesota. As a result operational services were available to local authorities during the Illinois River flood of December 1982 and the Pearl River flood of April 1983.

Satellite data have also been used to locate ice cover and ice dams on a number of North American rivers. The movement of river ice is important because it poses problems for hydroelectric power plants, bridge piers, and ship navigation. In addition, when it breaks up it can form ice dams which lead to severe flooding. This work has demonstrated that for rivers over 1 km wide, daily monitoring is possible and can produce useful flood warnings. For rivers in the far north, out of the view of geostationary satellites, orbiting satellites can be used to provide the same type of warning.

10

Measuring the oceans

'Before their eyes in sudden view appear
The secrets of the hoary deep, a dark
Illimitable Ocean, without bound,
Without dimension, where length, breadth and height,
And time and place are lost…'
John Milton 1608–74

OVER 70% of the Earth is covered by water. The role of the oceans in controlling the global climate has been appreciated by meteorologists for many years, but until the advent of satellites there was far too little information about conditions at sea. Meteorologists had to rely principally on weather ships situated mainly in the North Atlantic. While these were able to provide high-quality local observations there were vast areas of the ocean for which no data were available. Equally, many aspects of physical oceanography depended on limited measurements and so there were huge gaps in our knowledge of the behaviour of the oceans. Indeed, it is not much of an exaggeration to say that the description of the Gulf Stream and North Atlantic Drift had not advanced dramatically since Benjamin Franklin first published a remarkably realistic map in 1770.

In spite of the shortage of information, an increasing body of evidence, relying principally on shipboard observations, had been built up linking abnormal weather conditions with anomalous sea surface temperatures. By the late 1960s the British Meteorological Office was relying on temperature measurements south of Newfoundland, Canada, to assist in the production of monthly forecasts. The reasoning behind this was that when the water in this region was unusually warm, depressions moving into the North Atlantic were likely to be strong enough to push across northern Europe. Conversely, when water temperatures were low the depressions were less vigorous. So, low sea temperatures south of Newfoundland, it was argued, should lead to high pressure over Europe, while higher than normal sea temperatures should result in a more westerly pattern.

A similar analysis has been developed by Jerome Namias at the Scripps Oceanographic Institute, California, to explain the occurrence of cold and mild winters

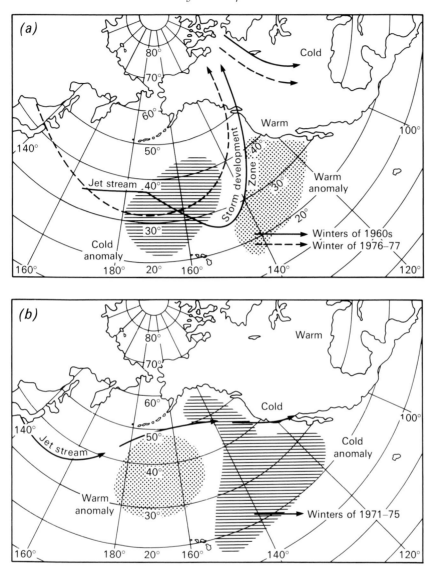

Fig. 10.1. The year-to-year influence of Pacific Ocean surface temperatures on the jet stream have been put forward as a possible explanation for differences in winter weather across North America: (a) two bad winters; (b) a mild winter.

in the eastern United States during the 1960s and early 1970s. He has suggested that when the central North Pacific is cold and the sea is warm off the west coast of America the jet stream swings south and then northward into Canada and so brings arctic air down across the eastern side of the continent. Conversely, when the central North Pacific is warm and the water is cold off the west coast the jet stream is diverted northward and swings south over California, thereby bringing mild air up across the eastern seaboard of the United States.

The availability of satellite measurements of sea surface temperature opened the way not only to the examination of these theories but also to the study of other aspects of the surface structure of the oceans. In addition the development of microwave techniques offers the prospect of wind and wave measurements. But to produce useful observations for practical weather forecasting is laborious. Indeed, thus far, the story of ocean measurements from satellites has been largely about the development of instrumentation and the evaluation of observations.

10.1   Sea surface temperatures

Since the launch of Nimbus 3 in 1969 it has been possible to extract some qualitative information about sea surface temperatures (SSTs), but obtaining absolute measurements has depended on a variety of factors. The usual problems of clouds and aerosols in the field of view of the infrared radiometers (see Chapters 4 and 7) had to be addressed. In addition, the measurements made by microwave radiometers required careful calibration because of the variation of emissivity with both frequency and wind speed (see Chapter 4). More importantly the very nature of what was being observed from both satellite and shipboard measurements required careful and exhaustive comparison since each technique measures different things. Most shipboard measurements have, as a matter of routine, been made at the engine intake. This is typically some 5–10 m below the surface. Over the years it has been established that this gives a bias of about 0.3–0.5 °C compared with the traditional oceanographic method of using a bucket to sample the surface water. More modern techniques now use towed thermistors which measure the top few centimetres, while instrumented buoys measure the top metre or so.

Microwave radiometers measure the temperature of the top few millimetres over an area some 150 km across, while infrared methods measure the temperature of the top few micrometres over an area a few kilometres across. By comparison shipboard measurements are few and far between and are restricted principally to the shipping routes of the Northern Hemisphere, so it is difficult to get adequate coverage compared with the satellite data.

The differences inherent in these contrasting measurement techniques are physically real. Not only do they include changes of temperature with depth and from place to place, but also with time of day. The diurnal effect is very important, as under sunny conditions the top metre or so can heat up by as much as 2 °C. This is comparable to the size of large-scale temperature anomalies, which are of so much interest to both weather forecasters and climatologists. Measurements have to be made at different times of the day to discriminate between diurnal effects and persistent abnormal temperatures.

Unravelling of these various effects has taken many years. The most complete observations have been obtained from four sets of satellite-borne instruments:

(i) the (AVHRR) instruments on board the TIROS-N/NOAA satellites which have provided day and night coverage over a 2700 km swathe using its 3.7, 11, and 12 μm windows to produce SSTs with a spatial resolution of 25 km;

(ii) the High Resolution Sounder (HIRS) and the microwave sounder unit (MSU) flown on TIROS-N/NOAA satellites using the infrared and microwave channels which sense radiation from the surface and lower atmosphere (see Chapter 4) have been used to produce SSTs with a spatial resolution of 125 km over a 2300 km swathe;

(iii) the Scanning Multichannel Microwave Radiometer (SMMR) on Nimbus 7, using channels at 6.6, 10.7, 18, and 21 GHz has produced measurements with a spatial resolution of 150 km over a 780 km wide area, but cannot produce useful information within 600 km of land;

(iv) Visible and Infrared Spin Stabilised Radiometer Atmospheric Sounders (VAS) on GOES 4, 5, and 6 satellites have been used to produce SSTs with a spatial resolution of 50 km over their field of view.

Analyses of these data produced from these sources have shown that individually they are capable of producing results that are accurate to within 1 °C. When the data are combined results are improved, but there are still systematic errors between the instruments due to stratospheric aerosols, water vapour and, to a lesser extent, cloud cover and winds. So while accuracies of better than 1 °C can be obtained, there is still considerable room for improvement. Where such absolute accuracy is not essential, important results have been obtained based on measuring temperature differences from place to place.

10.2 Ocean currents

One of the earliest successes of satellite SST measurements was to provide new insights into the surface structure of ocean currents. It has long been known that major currents like the Gulf Stream shed eddies of warm water or entrap rings of colder water. These rings are between 100 and 300 km wide and can retain their structure and identity for years. They rotate clockwise once every two to four days with current velocities of 0.5–1.0 m s^{-1}, and can be regarded as the ocean equivalent of atmospheric storms. Many of their details remained unknown. With satellite radiometers it is possible to measure their structure to an accuracy of less than 1 °C and with a spatial resolution of 1 km. This has enabled researchers to study the behaviour of these eddies and to gain a better insight into their role in heat transport in the oceans. At the same time satellite studies have plotted the meandering changes in the paths of the major currents and examined the sharp temperature differences ('fronts') that exist in the oceans.

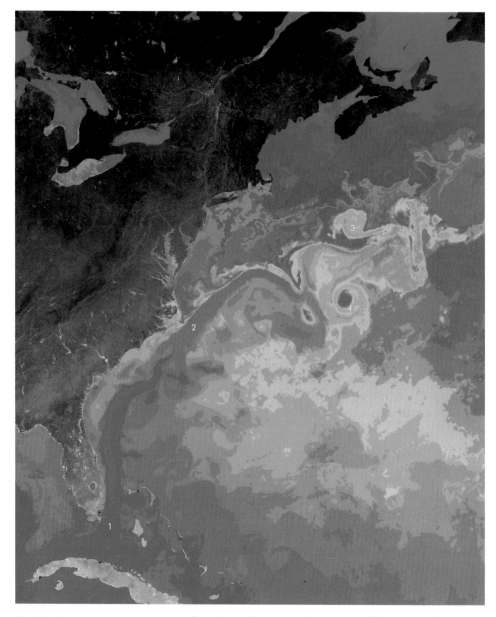

Fig. 10.2. Temperature measurements from 35 satellite passes during April 1984 were used to produce this image of the east coast of the United States and the Gulf Stream. It shows clearly warm water sweeping up from the Gulf of Mexico (1) and out across the North Atlantic (2) while cold water moves down from Newfoundland to Cape Cod. Of particular interest is the way in which the Gulf Stream tends to break up forming warm (3) and cold (4) eddies (with permission of NASA).

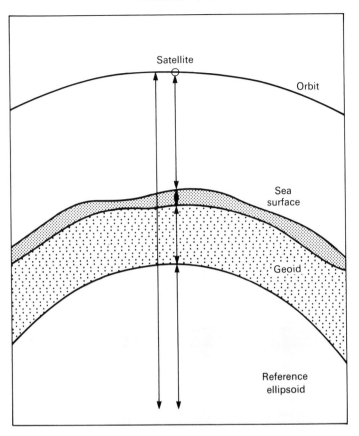

Fig. 10.3. If the orbit of a satellite is accurately known, it is possible to measure variations in the surface height of the oceans. To the extent that this surface differs from the geoid – the undisturbed shape defined solely by gravity – it is possible to draw conclusions about the large-scale movements of the oceans.

An even more important aspect of ocean currents is the amount of energy they transport. This requires a knowledge of not only their temperatures but also of how fast they flow. Velocity can be deduced from their influence on ocean topography. If the ocean were still, its large-scale surface topography would be that of the geoid – the surface that would be formed by the Earth's gravity field alone. This undisturbed shape is influenced solely by gravity; however, large-scale ocean movements cause bulges and depressions in the sea surface. The effect is small but measurable. For example there is a sea surface height difference of 1 m in a distance of 100 km across the Gulf Stream. The difference between the actual topography and the geoid gives a measurement of the speed and direction of the major ocean currents. As the currents vary with time, so does the sea surface height.

The first measurements from space of the strength of ocean currents were made by the radar altimeter on the experimental satellite GEOS 3 launched in 1975. This instrument was capable of measuring surface height with an accuracy of 30 cm, and with repeated observations achieved reproducible results to within 10 cm. Nearly three and a half years of continuous altimeter data were used to study seasonal and annual variations of the Gulf Stream. The results show that the dominant seasonal variability is between the strongest surface current in late winter and the weakest surface current in late autumn.

More accurate observations of ocean currents were made in 1978 by the altimeter onboard SEASAT. In the three months SEASAT functioned its altimeter demonstrated a precision of 5–7 cm and was used to make accurate measurements of fluctuations in the surface level of the oceans. In measuring the variability of the sea surface topography about the mean, it was possible to show that the greatest changes (10–25 cm) were associated with the strong Western Boundary Currents. These currents include the Gulf Stream, the Kuroshio Current, the Agulhas Current, and the Brazil–Falkland Confluence. Large variations also occur in the circumpolar Antarctic Current. Conversely, over most of the oceans there is surprisingly little variability. It was also possible to infer the velocity of the Gulf Stream from measurements of the slope across the edge of the current, and to model the flow that would lead to the observed cross-section. Values of up to 200 cm s^{-1} were produced. These are in reasonable agreement with surface observations.

More recently an improved altimeter was launched on GEOSAT in 1985. The spacecraft was specifically designed around this instrument to provide 18 months of ocean surface measurements. The altimeter was identical to that on SEASAT, but the longer observation period is expected to produce more accurate observations. These will include: measuring the surface to a precision of 3.5 cm in the presence of 2 m waves; estimating significant wave height to within ± 0.5 m, or $\pm 10\%$, whichever is greater; and measuring wind speeds to an accuracy of ± 1.8 m s^{-1} over the range 1–18 m s^{-1}.

10.3 The El Nino

While the role of SSTs in abnormal weather patterns has fascinated meteorologists for many years, it was events in the tropical Pacific in 1982 and 1983 which brought the topic into much sharper focus. This was because not only did much of the equatorial Pacific experience record-breaking temperatures, but the weather elsewhere in the tropics exhibited an extraordinary range of extremes. These included severe droughts in Australia, the Sahel, and southern Africa. All of these events were linked with a phenomenon known as the 'El Nino'. Originally the El Nino (Spanish for 'the child') referred to conditions observed during some years off the coast of Peru around Christmas, and hence were associated with the nativity. Warm water spread over the top of the normally cold nutrient-rich coastal waters, with catastrophic effects for the local anchoveta fisheries. Now the term has come to mean

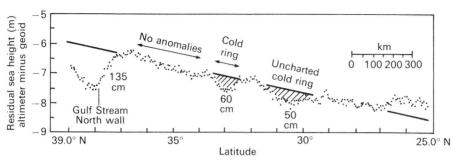

Fig. 10.4. An example of a satellite altimeter measurement of the sea surface, showing the height variation across the Gulf Stream.

occasions when anomalously warm surface waters cover much of the tropical Pacific.

The importance of such events is that they are part of large-scale atmospheric and oceanic fluctuations which appear on average every four years, but the time between each El Nino can vary from two to ten years. The broad weather patterns linked with these events have been recognised since the 1920s when Sir Gilbert Walker presented evidence that 'when the pressure is high in the Pacific Ocean it tends to be low in the Indian Ocean from Africa to Australia and vice-versa'. Known as the 'Southern Oscillation', this behaviour is now seen as an integral part of the process of the El Nino. Indeed, meteorologists often refer to the El Nino Southern Oscillation (ENSO) as a single phenomenon.

What happens is that in a remarkably predictable way the warm water, which first appears off the coast of Peru can, over the course of a year, spread like a huge tongue up to 8000 km across the equatorial Pacific. At the same time, atmospheric pressure declines over the eastern Pacific and rises over Australia and the Indian Ocean. With this, the surface winds along the equator reverse. Where the pressure drops the rainfall rises dramatically – in Guayaquil, Ecuador, the excess rainfall between October 1982 and June 1983 was nearly 3 m. In the high pressure region drought conditions prevailed.

The consequence of these shifts is to alter the pattern of intense convective activity which girdles the Earth close to the equator (the intertropical convergence zone – ITCZ), and with it, rainfall patterns around the world. In 1982 the ENSO was associated with not only the drought in Australia, but also with the delay in the monsoon in India and droughts in northeast Brazil, Central America, and southern Africa. Moreover, the effects of these unusual patterns are not confined to the tropics. Earlier ENSOs have been implicated with the extraordinary cold winter of January 1977 in the eastern US and the string of weather disasters of the summer of 1972 including the failure of the Russian harvest.

This gives the impression that there is a simple connection between these events. Unfortunately this is not the case because, in spite of there being a standard descrip-

Fig. 10.5. Temperature measurements of the Pacific Ocean show the effect of the El Nino in 1982–83. The top image (20 January 1984) shows the normal distribution with the warmest water (1) in the western equatorial Pacific and a tongue of cooler water (2) stretching out from South America. In the middle the effect of the El Nino is seen on 20 January 1983 in the far greater extent of warm water and the disappearance of the cooler equatorial tongue. The bottom image shows the temperature difference between these two sets of observations with the marked warming of the eastern equatorial Pacific (3) and a corresponding cooling of the western tropical Pacific (4) (with permission of NASA).

Fig. 10.6. Measurements of ocean colour in early May 1981 by Coastal Zone Color Scanner (CZCS) on Nimbus 7, showing the east coast of United States and variations in biological activity in western North Atlantic. The areas rich in phytoplankton are shown in red and those of low productivity in blue and purple. The formation of warm eddies of low-productivity water in the Gulf Stream can be seen at 1, 2, and 3 (with permission of NASA).

tion of an ENSO event, there are important differences from occasion to occasion. For instance the event which began in May 1982 appeared first in the western Pacific and worked its way back toward South America. This was in spite of the fact that the Southern Oscillation was at a record extreme. More importantly, in some years when conditions appear ripe for an ENSO, it inexplicably fails to materialise. The only way to unravel these complexities is to collect more detailed observations of future events. This will depend heavily on satellites, not only to measure SSTs but also to observe the atmospheric patterns that are integral to the process. In particular the way in which successive atmospheric disturbances (both tropical storms and less organised convective activity) move along either side of the equator may prove to be an essential part of the build-up of an ENSO.

The contribution of satellites is seen not only in the direct measurement of SSTs but also in wider observations of global processes of which regional anomalies may be but just one part. This has become clear with the measurements that were made during the 1982–83 ENSO and in subsequent years. Notably, parallel observations of the tropical Atlantic showed a pattern of compensating though less extreme temperature anomalies. When the Pacific was abnormally warm in 1983, the Atlantic was slightly below normal. In 1984 the scene reversed with the Pacific cooling and the Atlantic being markedly warmer. This is the first time this see-saw effect has been measured with any precision. It suggests that simultaneous measurements of all the tropical oceans must be obtained if we are to understand and predict not only the El Nino but also the role of variations in the tropical oceans in producing abnormal weather patterns around the globe.

10.4   Ocean colour measurement

The colour of the oceans, as seen from space, is a direct measure of the biological productivity of the surface waters. Although this parameter is not of immediate meteorological interest it is of considerable oceanographic importance and potential climatological significance. The amount of light at different visible wavelengths reflected by the top few metres of the ocean is strongly dependent on the amount of phytoplankton and sediment present. Most deep ocean waters contain no phytoplankton or sediment. They exhibit the properties of pure water and appear deep blue. But on the continental shelves, or in the cold currents bringing nutrient-rich waters from high latitudes, or where cold deep water is upwelling, the colour is much greener due to chlorophyll in the phytoplankton.

The colour of the oceans was measured for nearly eight years by the Coastal Zone Colour Scanner (CZCS) on Nimbus 7. This radiometer had a spatial resolution of 0.8 km and scanned a swathe 1600 km wide. It measured the colour of the oceans in four separate narrow visible wavelength bands. From these observations, the difference in the amount of blue and green light observed was used to calculate chlorophyll concentration. This approach is similar to the calculation of the global vegetation index on land (see Section 9.1).

The most obvious long-term meteorological implication of this type of measurement is that it will provide a better understanding of the role played by ocean productivity in the uptake of CO_2 from the atmosphere. This process is crucial in determining how rapidly CO_2 emitted by fossil fuel combustion accumulates in the atmosphere. In the shorter term these measurements have already provided complementary observations of the behaviour of ocean currents, and have proven to be of commercial value in guiding fishing fleets to nutrient-rich areas.

10.5 Fuelling hurricanes

The way in which hurricanes pick up energy from the oceans is an essential factor in understanding their behaviour and improving forecasts of their motion, but, until recently it has not been possible to measure how much energy is transferred from the oceans to the atmosphere. Now analysis of satellite observations of the SSTs before and after the passage of a hurricane have provided the first indications of the scale of the processes involved. They show that with all hurricanes there is some sea surface cooling, but large and persistent changes occur only with the strongest storms. Cooling in excess of 3 °C over the open ocean and lasting more than two weeks has been observed with vigorous hurricanes. At higher latitudes and closer to shore even greater cooling may occur where vertical mixing introduces colder deep water. The amount of cooling and its persistence is correlated with the strength of the storm.

The process of cooling is complicated by the speed at which the storm moves. Slow-moving systems tend to cool close to the centre of the storm-track while more mobile systems cool most markedly as much as 70 km to the right of the path of the hurricane. More importantly there is evidence to support the long-standing theory that hurricanes gravitate toward warmer water.

10.6 Winds and waves

The use of microwave radiometers, scatterometers, and altimeters enables scientists to measure the surface roughness of the oceans (see Chapter 4). While these techniques are still at an early stage of development, they have already produced results which demonstrate their potential value. Observations from GEOS 3, SEASAT, and Nimbus 7 have been used to obtain useful values for wind speed and direction, and wave height. Understanding these results has been difficult because of the range of processes at work. As explained in Chapter 4 the microwave emissivity and scattering properties of the sea surface are complicated functions of temperature and wave structure. The form of the waves in turn is the product of both local weather conditions and past events. Only after lengthy calibration studies of the equipment under operational conditions has it been possible to produce reliable results.

So far oceanographers have had considerable success in exploiting microwave altimeters, radiometers, and in particular the scatterometer on SEASAT to measure wind and waves. The advantage of the scatterometer (see Section 4.8) is that it can measure the amount of emitted polarised microwave radiation that is scattered back to the satellite. This provides much more data about surface conditions than can be obtained from radiometric observations. At a given angle to the flight path of the satellite the amount of back-scatter depends on two factors – the size of the ripples and their orientation with respect to the plane of polarisation of the pulse

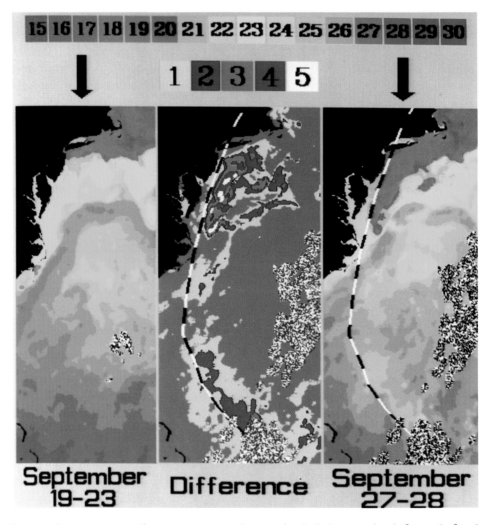

Fig. 10.7. *Changes in sea surface temperature can be seen clearly in images taken before and after the passage of hurricane Gloria up the western North Atlantic. On the left a composite of satellite temperature measurements from 19 to 23 September 1985 and on the right the composite for 27 and 28 September showing the path of the hurricane. In the centre is the difference in temperature before and after the hurricane showing that the greatest cooling occurs to the right of the storm's track (with permission of University of Rhode Island).*

Fig. 10.8. The winds over the Pacific measured by SEASAT between 6 and 8 September 1978. The strength of the winds is shown by both the length of the white arrows and the background colour coding (light winds $< 4\,m\,s^{-1}$ are in light blue, through to strong winds $> 14\,m\,s^{-1}$ in yellow). The strongest winds are in the 'roaring forties' of the Southern Hemisphere, especially associated with two intense depressions to the east of New Zealand (with permission of NASA).

of radiation transmitted by the scatterometer. The first is dependent on wind stress and hence wind speed at the surface, while the second is related to wind direction. The returning signal can be converted, by means of a complicated mathematical algorithm, to provide an estimate of wind speed and direction. This process requires measurements at different wavelengths and with both vertically- and horizontally polarised radiation. SEASAT measured wind speeds with an accuracy of $\pm 1.8\,\mathrm{m\,s^{-1}}$ over the range 3–28 m s^{-1} and wind direction with an accuracy of $\pm 20\%$. These observations showed that a single orbiting scatterometer could produce a five-fold increase in the number of wind measurements recorded at sea in the Northern Hemisphere and a 60-fold increase in the Southern Hemisphere, compared with existing shipboard observations.

In obtaining improved estimates of wind speed and direction much progress has been made in understanding the nature of air–sea interactions. Not only is roughness dependent on wind speed and the temperature difference between the water and the atmosphere, but it is also affected by weather conditions across the field of view of the scatterometer. Because this is some 150 km across it can contain a lot of weather. This is particularly important where cold air is streaming across warm water and intense convective activity occurs which can lead to large differences in wind speeds over short distances. Observations are further complicated by heavy rainfall which affects the surface roughness. It will require a resolution of 25 km to make useful measurements on standard weather systems. Even then it will not be possible to resolve the fine structure of many wind patterns.

The scatterometer data from SEASAT have been evaluated to test the potential for using microwave instruments in weather forecasting. While the level of agreement between wind speeds observed on board ships and those obtained from SEASAT data is reassuring, it is obvious that both the measurement technique and the spatial resolution will need to be improved if useful predictions of storms at sea are to be made by satellite. Part of this progress may come from making complementary measurements using other microwave equipment. There is, however, a more pressing reason for using other forms of measurement. SEASAT was an experimental system which failed after only three months and also, as there were delays in subsequent programmes, there will be a gap of more than ten years before another scatterometer is put into operation.

Both radiometers and altimeters can be used to derive information about wind and waves. In particular, radiometers can detect not only the effect of capillary waves, but also the tilting of the larger waves and white-capping. So far it has been demonstrated that results from these instruments can effectively complement scatterometer observations. More than six years of data have been obtained from the scanning multichannel microwave radiometer (SMMR) on Nimbus 7, which can measure wind speed to an accuracy of $\pm 2.0 \mathrm{~m~s^{-1}}$ over the range 0–20 m s^{-1}. However, the SMMR cannot provide information about wind direction.

At an earlier stage it had been demonstrated that more general climatic results of wave height and hence wind speeds could be extracted from altimeter data obtained from GEOS 3 over a period of four years in the late 1970s. Observations made from December to March and June to September show many of the features known to mariners – the mid-latitude westerlies, the horse latitudes, the trade winds, and the doldrums. They confirm interesting variations between the seasons. Whereas the wave heights associated with the westerlies in the Northern Hemisphere decline by more than a factor of two between winter and summer, those in the Southern Hemisphere (in the 'roaring forties' and the 'screaming fifties') continue unabated, save to move back and forth over about 10° of latitude.

10.7 Synthetic aperture radar

So far the application of synthetic aperture radar (SAR) (see Section 4.8) to weather-related aspects of oceanography has not advanced beyond the experimental phase. The high resolution instruments on SEASAT and the Space Shuttle have obtained detailed images of wave patterns. These show that wave spacing and the direction of wavefronts can be measured in all weather conditions. Observations have been made of wave structure beneath a mature hurricane. These offer the prospect of improved modelling of both conditions in tropical storms and a better understanding of the air–sea interactions involved. Simultaneous aircraft and satellite measurements are also being used to study the generation of extreme waves by extratropical storms. The objective of this work is to produce better forecasts of the behaviour of these waves with a view to reducing the damage they cause when they make landfall or strike offshore installations.

11

Snow and ice

'Great God! This is an awful place'
Captain Robert Scott
(describing the South Pole)

THE AREA of the Earth routinely covered by snow and ice, often termed the 'cryosphere', is an essential element of the global climate. The most permanent components are the ice-caps of Greenland and Antarctica and the glaciers and permanent snow cover of mountain ranges. Sea ice covers some $2.0 \times 10^7 \text{ km}^2$ of the ocean surface. More importantly it exhibits huge seasonal variations which exert a major influence on the weather and climate, locally and globally. Even more variable is snow cover in the Northern Hemisphere. In mid-winter it extends over some $4.5 \times 10^7 \text{ km}^2$, but covers only the permanent ice-caps and glaciers in late summer. Moreover, it can fluctuate rapidly over a few days as weather patterns change.

The combined effect of all of these fluctuations is to establish an annual pulse, with the extent of snow and ice expanding and contracting in each hemisphere. In the Northern Hemisphere this pulse is the product of the combined effects of continental snow and sea ice. In the Southern Hemisphere the changes are dominated by the expansion and contraction of the pack ice around Antarctica. While the effects in the two hemispheres are out of phase and hence tend to balance one another out, changes in distribution from week to week are important in weather forecasting. Longer-term changes over the years may prove to be crucial in understanding climatic change.

The high albedo of snow and ice, and the high infrared emissivity of snow are of great meteorological consequence. Each in its way plays a vital role in both influencing climatic behaviour. The high reflectivity of snow and ice means that it reduces the amount of solar energy absorbed at the surface during the day. At night high emissivity leads to rapid cooling and is a major factor in producing very low temperatures in winter (see Section 7.3). Because snow and ice are relatively

134

easy to see from space, some of the earliest integrated measurements were made of the seasonal changes in their extent. These were of limited coverage as they could only be made in sunlit areas, which excluded much of the polar regions during their respective winters. While integrating visible images enabled scientists to distinguish between continuous snow and ice cover and variable cloudiness, the technique had limitations. The margins of snow and ice areas where the cover is patchy make it hard to distinguish between snow and ice, and the underlying surface. Also such areas are regions of high cloudiness, especially over the oceans. This undermines the accuracy of measurements using visible images. In particular, in the stormy almost totally cloud-covered oceans around Antarctica, discrimination between pack ice and cloud requires microwave techniques.

11.1   Visible and infrared measurements

Despite the limitations, early measurements provided valuable information for the routing of ships in both arctic and antarctic waters. On land, they were used to estimate the extent, water equivalent, and the onset of spring thaw in the western US. From these measurements it was possible to make estimates of run-off, and from these to plan irrigation and flood control strategies.

The measurement of snow and ice cover is now routinely used to make considerable economic savings. On the Great Lakes it costs around $1 million per day when shipping is stopped by ice cover. Prior to satellites the Lakes used to be shut for about two months per year. However, in the exceptionally severe winter of 1977 only one month was lost and in 1978, when the weather was nearly as bad, they were never totally closed. Satellites have played a significant part in this saving, but in recent years shore-based synthetic aperture radar has also been a major factor in this success.

In the western US, where 70% of total run-off is from the winter snow-pack, satellite mapping is a major aid to the management of the hydrology of the region. Using satellite pictures is 200 times cheaper than making aerial surveys, and now over 600 river basins are mapped each winter. It is estimated that run-off prediction has been improved by 6–10% and the consequent value to irrigation and hydroelectricity is approximately $30–40 million per year.

11.2   Records of snow cover

Variations in the extent of snow cover in the Northern Hemisphere have been measured since the late 1960s. Weekly charts have been prepared in the United States. These show the areal extent and brightness of continental snow cover but do not indicate snow depth. From these data estimates can be made of how snow cover varies from the climatic normal and these are used, over time, to build up a better picture of year-to-year variations and long-term trends in the extent of cover. In addition, long-term series have been prepared of the changes in the seasonal extent of snow which may eventually be linked to other climatic changes.

Weather forecasters and climatologists soon found these maps of potential value to understanding fluctuations in the weather. These measurements gained particular prominence in 1974 when there was widespread comment about the possibility of sudden changes in snow cover triggering another Ice Age. Estimates of the annual average area covered by snow in the Northern Hemisphere showed a dramatic increase in 1972 which was sustained over the following year. This shift was seen as an example of a process that could lead to a sharp cooling of the global climate. Subsequent measurements showed that this was a temporary effect and that the extent of continental winter snow cover varies considerably from year to year, especially over central Asia, but it did stimulate a great deal of scientific interest in possible mechanisms of climatic change.

The possibility of anomalous snow cover affecting the weather has also been proposed by climatologists as an explanation of shorter-term sustained abnormal weather. There is some evidence that both positive and negative anomalies in the snow cover of northwest America, the Baltic, the Caspian and Caucasus, the Tibetan Plateau and Mongolia–North Korea occur synchronously. In specific years these

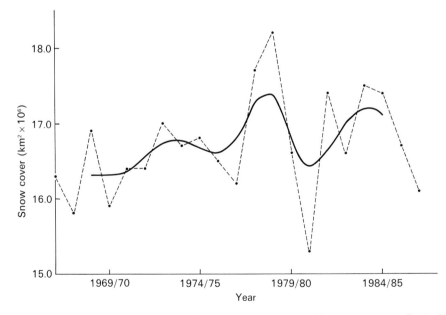

Fig. 11.1 North American winter snow cover from 1966–67 to 1986–87. The annual values (dashed line) are for December to February, inclusive, and the longer-term trend (solid curve) is shown by the five-year binomial running mean.

136

abnormal patterns may contribute to exceptional weather conditions. For example, the extreme winter of 1962–63 in Europe – the coldest since 1830 – may have been prolonged by widespread and long-lasting snow cover. More recently, an analysis of the cold winter of 1983–84 in the US provided important clues in support of this theory. December 1983 was exceptionally cold over much of the United States, breaking many long-standing records. It produced unusually extensive and deep snow cover over the eastern half of the country. Satellite data showed that throughout the month snow covered an area 1×10^6 km^2 more than in an average December. The effect of this seems to have been to prolong the wintery weather appreciably despite the broad weather patterns shifting to what would normally be milder conditions. During January 1984 it was estimated that in parts of the mid-West day-time maximum temperature was 5 °C lower than would have been expected on the basis of prevailing atmospheric conditions. In addition the amount of precipitation was reduced. This suggests that the differences were due to the high reflectivity of the snow which both reduced heating near the ground and cut down the convective activity in the atmosphere which often produces rain and snow.

In the Southern Hemisphere snow cover in South Africa, Australia, and New Zealand is either too infrequent or spatially insignificant to warrant studying from space, but in South America large changes occur from year to year. While these are not of global significance they may exert interesting influences on the regional climate and so are measured on a routine basis by satellites.

11.3   Microwave measurements of sea ice

With the launch of microwave radiometers on Nimbus 5 in 1972 and Nimbus 6 in 1975, the quality of the observations of sea ice improved dramatically. This was because in the microwave region the interference of clouds was largely removed, and the contrast between the emissivity of ice and water was high. This enabled areas of ice cover and seawater to be distinguished, and in areas where there was a mixture of ice floes and open water the sea-ice concentration could be estimated. Smaller differences in the emissivity of snow cover and uncovered land meant that observations of the extent of continental snow were improved to the same extent. Subsequent work has, however, demonstrated that useful observations of continental snow cover can be made using multifrequency microwave radiometers.

The measurements made from Nimbus 5 enabled a comprehensive picture of the seasonal changes of arctic and antarctic pack ice to be produced. These results are particularly interesting as they have been used to produce false-colour images which portray annual climatological changes in graphic detail. For example, around Antarctica the extremes of the extent of the ice can be seen in the images for August 1974 and February 1975. The antarctic land mass is overlaid in black, with the tip of South America in the upper left quadrant. High ice concentrations near the coast

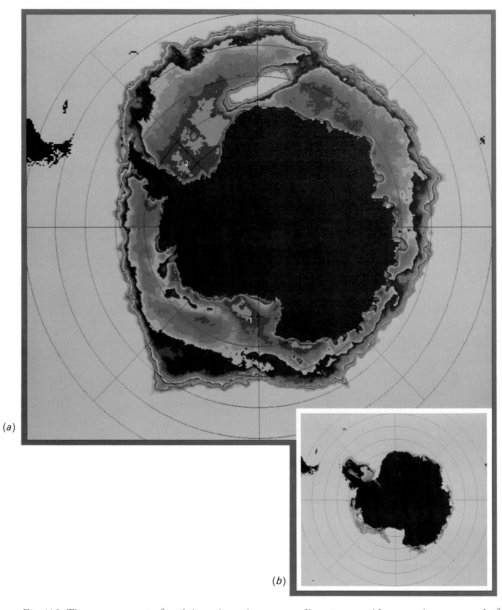

(a)

(b)

Fig. 11.2. *The measurement of pack ice using microwave radiometers provides a continuous record of changes in sea ice in polar regions throughout the year. This pair of images shows the maximum extent of ice around Antarctica in* (a) *August 1974 and at its minimum extent in* (b) *February 1975 (with permission of NASA).*

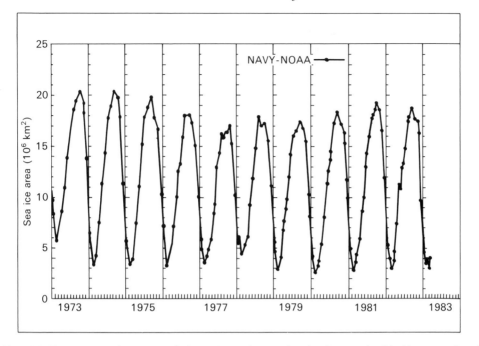

Fig. 11.3. Variations in the extent of Antarctic sea ice can be clearly seen in this 10-year series of microwave radiometer measurements (from H. J. Zwally, Annals of Glaciology, *5, 191 (1983)).*

are shown in shades of red and low values near the ice margin in shades of blue. The winter image (August 1974) shows the full ice cover surrounding Antarctica extending for more than 1000 km from land into the Ross and Weddell Seas. The large ice-free area in the eastern part of the Weddell Sea is known as a 'polynya'. Although polynyas are found in many areas, even in the depths of winter, the Weddell example is of particular interest. It does not form every year, but when it does it is found in approximately the same position. The air–sea interactions that lead to its persistence are not fully understood, but while wind action may initiate its formation, its survival clearly involves an upwelling of warm ocean water. It thus represents a major area of heat transfer from the ocean to the atmosphere.

By late summer (February 1975), the sustained heating of continuous daylight has melted more than 80% of the sea ice. All that is left is an area of ice off the Wallgreen coast and one that is protected from the strong westerly winds in the lee of the Antarctic Peninsula. These huge seasonal expansions and contractions of ice are a major controlling factor in the annual shifts of the weather patterns over much of the Southern Hemisphere. Long-term changes in the extent of the ice cover may lead to alterations in mid-latitude storm-tracks and so provide an early warning of significant climatic variations. During the period 1973–83, the ice cover showed significant fluctuations, with the winter maximum decreasing appreciably by 1977 and returning earlier values over the next five years, but overall it is too early to make any definite observations about long-term trends.

In the Arctic, ice cover is less homogeneous and as a consequence the early microwave radiometers on Nimbus 5 and 6 were not able to obtain such accurate pictures of seasonal change. However, with the launch of the scanning multichannel microwave radiometer (SMMR) on Nimbus 7 in October 1978, it was possible to measure the emissivity of five frequencies and make much more precise measurements. The results for April and September 1979 show just what this instrument was capable of doing. These images show the ice cover in shades of purple, red, and yellow, with maximum ice concentrations in the regions of darkest purple. Light blue areas are open ocean. The scattered dark blue patches over the ocean signify heavy clouds or rain. The dark spot over the North Pole was not traversed by Nimbus 7. The image for April 1979 shows the ice near its maximum extent, covering some $1.5 \times 10^7 \, \text{km}^2$, following the dark winter months. It fills most of the Arctic Ocean, Hudson Bay, and the Sea of Okhotsk. It also spills out into the Bering Sea and drifts down the coast of Greenland. The narrow tongue of ice down the coast of Labrador shows how this cold southward-moving current is constrained by warm water from the Gulf Stream spreading northward. The influence of this water extends across the Atlantic to the northern coast of Norway. East of Greenland there is a large area of ice that has broken away from the main pack. By September 1979 the area of ice had shrunk to about half of its winter maximum and was confined to the central part of the Arctic basin.

The changes, although striking, are not as great as in the Antarctic. For while the maximum area of sea ice in both polar regions is about the same, the antarctic ice shrinks to only one-fifth of this area in summer. This contrast is a result of the marked difference in the character of the two polar oceans. Many rivers flow into the Arctic, producing a near-surface layer of comparatively fresh cold water which, because of its low density, inhibits the convection of warmer saltier water from below. This makes it easier for ice to form in the winter and also prevents it from melting from below because of the stable layer of low salinity water beneath it. In contrast, there is negligible fresh water run-off from Antarctica and therefore little to prevent deep, warm water from rising to the surface. This accelerates melting in the summer, but in winter persistent strong winds caused by intensely cold air draining off Antarctica push newly formed ice away from the coast. This produces much more ice than would occur if the winds showed the more variable patterns that exist at lower latitudes.

11.4   Variations in sea ice

Microwave radiometers provide an excellent picture of broad seasonal changes in polar sea ice, but there are many other features of these variations that require more detailed examination. Except on the Filchner and Ross Ice Shelves off the coast of Antarctica polar sea ice is at most a few metres thick. But, because it extends over hundreds of kilometres it is subjected to large forces due principally to surface winds and to a lesser extent ocean currents. These lead to the ice breaking up to form ridges and humps and to expose open water. Their effects are of considerable interest to meteorologists. The patterns of ice motion reveal a lot about the forces at work. The extent of open water is of more direct interest as sea ice is a very effective insulator, reducing heat loss from ocean to atmosphere to 1% of the loss in ice-free regions.

The most direct way to refine the details of what is measured by microwave radiometers is to make simultaneous surface measurements. Considerable work has been carried out in both the Arctic and Antarctica to do this. It shows that the emissivity of the ice is dependent on its age, history of formation, and surface characteristics. First-year ice has a relatively high salinity and hence is highly emissive irrespective of the microwave frequency of the observations, but snow cover reduces the emissivity. Conversely, melting in spring and summer increases emissivity through the formation of slush and puddles on the ice. This leads to considerable problems in measuring ice concentrations at these times of year, when air temperatures fluctuating about the freezing point can lead to significant changes in emissivity but little change in ice cover. Overall, it is concluded that the ice edge can be detected with an accuracy of 10–45 km. While this method lacks the resolution of optical and infrared observations, it is less sensitive to atmospheric and oceanographic effects.

Optical and infrared techniques can come into their own in studying the motion of pack ice. Images obtained using high resolution radiometers can provide useful information about the proportion of open water and the general pattern of ice movement. If pictures can be obtained within a few days of each other then it is possible to make more accurate estimates of the movement of the ice by tracking the major ice floes or icebergs. However, these efforts are often frustrated by the frequent cloud-cover in polar regions and are restricted to the infrared during the winter when the polar oceans are in continuous darkness.

The only way to get detailed pictures of sea ice in all weather conditions is by using synthetic aperture radar (SAR) (see Section 4.8). Capable of spatial resolution of about 25 m, this technique was successfully demonstrated in SEASAT. Bright regions in a SAR image are the areas that reflect energy back toward the spacecraft and generally have surface roughness characteristics with a scale of a few centimetres. This tends to be old and ridged ice and the wind-ruffled oceans. Smooth horizontal surfaces such as calm water and new ice reflect energy away and appear black in the images. It is now possible to discriminate between various ice types and water conditions and so make good estimates of how open water areas and ice formation change with time. Moreover, the positions of identifiable ice floes provide good measures of motion over time. Ice motion data are important for at least three

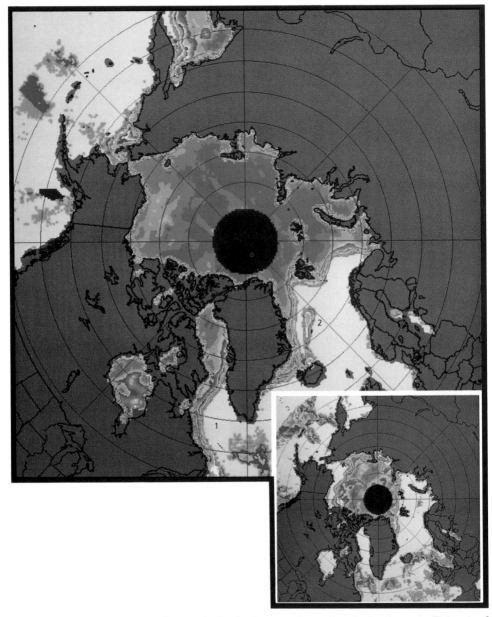

Fig. 11.4. Changes in the extent of sea ice in the Arctic are as dramatic as in the Antarctic. This pair of images shows the changes from late winter (top–April 1979) to late summer (bottom–September 1979) (with permission of NASA).

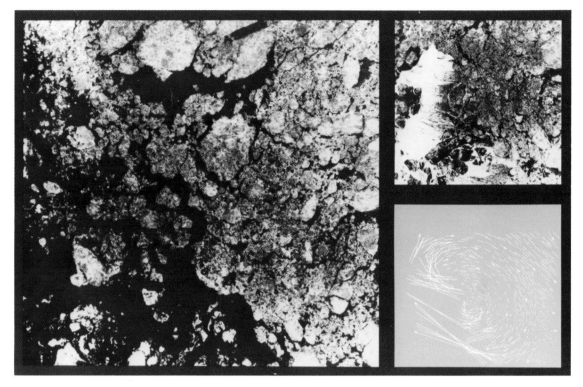

Fig. 11.5. Synthetic aperture radar can be used to plot the motion of pack ice. The images were obtained by Seasat on 5 and 8 October 1978 (on the top left and bottom, respectively) for a 100-km² region of the Beaufort Sea, north of Alaska. The motion of individual lumps of ice can be measured and the diagram in the top right shows how the pieces of ice moved relative to one another over the period of three days, but not the overall movement of the ice as a whole (with permission of NASA).

reasons. First, both the absolute and relative motion of the ice provide an estimate of surface current direction and velocity. Second, the relative motion gives a measure of the opening and closing of the pack ice and the rate at which open water areas are formed. Finally, these observations have important operational applications for routing ships and, in the Arctic, for mining and drilling operations.

11.5 Sea ice and the weather

Variations in the nature and extent of sea ice in polar regions influence the weather both locally and over a much wider area. For this reason, variations from year to year are of interest to weather forecasters. It has long been recognised that variations in the extent of pack ice in the North Atlantic could play a significant part in abnormal weather in northern Europe. For instance, an exceptionally cool June in 1972 in northwest Europe was widely linked with a higher than average number of icebergs in the northern Atlantic during the spring of that year. Similarly, the early formation of extensive pack ice off eastern Greenland in the late autumn of 1981 seems to have been a contributory factor to record low temperatures in Britain during the cold spells of December 1981 and January 1982 (see Section 7.3). More generally, the frequent cold winters of the late seventeenth and late eighteenth centuries have been directly linked with extensive pack ice north of Iceland.

11.6   Altimeter measurement of ice-caps

The volume of the Greenland and Antarctic ice-caps is a vital factor in the global climate balance. Any appreciable changes in the amount of water locked up in these ice-caps could signal important shifts in the climate, but, despite 25 years of intensive field work on the ground, it was not known whether the Greenland and Antarctic ice sheets are growing or shrinking. In Greenland there have been repeated measurements of one levelling traverse across the ice sheet. Results indicate a thinning of the ice at low levels, with a thickening further up the slope, but these trends are unlikely to be representative of the entire ice-cap. In Antarctica there has not been a levelled transverse and the surface topography in most areas is poorly known. Conventional survey techniques, even airborne measurements, will never provide data at a sufficiently high spatial density to cover the entire ice sheet. The only way forward is satellite altimetry.

The radar altimeter on SEASAT obtained almost one million useful measurements of surface elevation over Greenland and Antarctica. The measurement quality was within 0.1 m over the smoother portions of the ice sheets with increasing errors over sloping and undulating surfaces. At a more detailed level the SEASAT data have detected ice cliffs at the edges of the antarctic ice shelves with an accuracy of 0.1–1.0 km, so in the future it will be possible to measure changes in their positions and hence the rate of calving of icebergs. Indeed, the GEOSAT, launched in March 1985, has produced additional information of this nature.

This work is just beginning to bear fruit. Results published in December 1989 suggest that the Greenland ice-sheet has thickened 23 cm a year since the late 1970s. Observations from GEOS 3 between April 1975 and June 1978, from SEASAT in 1978 and from GEOSAT since 1985 have produced this first evidence of changing thickness. Comparable figures have not yet been obtained for Antarctica as the rate of accumulation of snow is much slower than over Greenland. But reconciling these observations with the observed current 2.4 mm per year rise in global sea levels suggests that the antarctic ice-sheet is ablating faster than the accretion over Greenland. Clearly more measurements are needed, and in this context SAR observations of the surrounding ocean which will distinguish between pack ice and icebergs and produce an inventory of the latter, will help to provide more data about the rate of discharge from the ice-cap.

12

Better weather forecasting

'I've looked at clouds from both sides now
From up and down as still somehow
It's clouds illusions I recall
I really don't know clouds at all.'
                                Joni Mitchell

MANY PEOPLE continue to believe that despite huge advances in technology there has been little progress in the quality of published weather forecasts. However, those who take a close interest in meteorology put a totally different interpretation on recent forecasting developments. While there are still striking examples of forecasters getting things splendidly wrong, the overall position is one of steady and, in places, dramatic advance. Satellite observations have played a significant part in the progress, but the most important contribution has been from computer science.

12.1   Bigger computers

The potential to use mathematical equations representing well-established physical laws to predict the behaviour of the atmosphere was first proposed by the British meteorologist Lewis Richardson in the early 1920s, but at the time this was impractical. As Richardson noted he 'played with a fantasy' of a forecast factory of 64 000 mathematicians, which he estimated would be the number needed to keep pace with the weather. This dream did not become a practical proposition until 1950 with the advent of high-speed digital computers. Even then the initial work was primitive and could not be used for forecasting, but once the principle had been established progress was rapid thanks to the increase in the speed of computers. Just how rapidly computing power has developed in recent decades can be gauged by the experiences of the British Meteorological Office. In 1959 it bought its first computer, a Ferranti Mercury, which could perform 3000 multiplications per second. In 1982 it installed a Control Data Cyber 205 capable of 4×10^8 multiplications per second. This computing power is needed to handle the huge amount of input data, much of it from satellites, and also to enable a detailed model of the atmosphere to be produced. For example, the Meteorological Office's model treats the atmosphere as 15 vertical layers over a grid with a horizontal resolution of 150 km – one-third of a million points in all. To run the model requires some 10^9 operations for each 12-min step in the forecast. A set of forecasts up to one week ahead, therefore, involve nearly 10^{12} calculations. By 1989 the installation of a new computer (the CDC ETA 10) offered an order of magnitude more speed and the prospect of yet more detailed models.

12.2   Problem areas

Progress in weather forecasting covers a wide variety of developments. These include improved physical understanding of atmospheric processes on all scales and improved measurement techniques. In all areas the role of satellite observations is of growing importance. The way these observations help to overcome forecasting problems varies from area to area. Broadly speaking, the problems facing weather forecasters can be divided into three main time scales. First, there is a need to improve the detail and precision of forecasts up to 24 h ahead. Second, there is the general question of just how far forward useful day-to-day forecasts can be extended beyond their current span of six or seven days. Third, there is the possibility of developing worthwhile by-and-large forecasts over periods of weeks and even months ahead.

The problems of short-range forecasting are easily defined. The British Meteorological Office has prepared an estimate of the major errors in its forecasts up to 24 h ahead (Table 12.1) during the years 1981–83. While the sources of errors in forecasts in other parts of the world may differ in detail, the broad conclusion is beyond dispute. The greatest immediate problem is in measuring the timing and intensity of rainfall or snowfall. Indeed, this difficulty may explain why so many people are sceptical about claimed improvements in the quality of the predictions. It only takes one unexpected drenching to completely undermine a string of accurate forecasts. For example, the failure in October 1987 to provide any warning of the worst storm to hit southern England in nearly 300 years led to widespread criticism of the work of the Meteorological Office. Avoiding such embarrassing mistakes requires continual and improved monitoring. Geostationary satellites, with their ability to obtain half-hourly pictures, are an obvious means of continually updating observations and refining short-term forecasts.

On a scale of a week or two the problem facing meteorologists of how far ahead they can make useful predictions depends on various factors. Theoretical studies suggest that extending the limits will require more detailed computer models and hence faster computers, and also much improved measurements of the initial state of the atmosphere. The only way to get these improved measurements is by using more accurate satellite systems. Longer-term predictions could depend on building up better knowledge of how the slowly varying components of the climate (eg.,

146

Table 12.1. *Causes of major errors in weather forecasts by the UK Meteorological Office* 1981–83[a]

Type of error	Number of cases	% of occasions
Rain – timing error 6 h or more	109	5.0
Rain – seriously underestimated	76	3.5
Widespread showers/thunderstorms underestimated	50	2.3
Temperature – major error	27	1.2
Widespread showers/thunderstorms overestimated	18	0.8
Rain – seriously overestimated	17	0.8
Snow – seriously underestimated	17	0.8
Severe gales underestimated	12	0.5

[a] This table only lists types of error which occurred 10 times or more.

sea surface temperatures, the extent of polar pack ice, and continental snow cover) can influence the formation of persistent weather patterns. Satellites have already demonstrated their capacity to measure these climatically important parameters. Meteorologists must now determine whether changes in these parameters actually lead to anomalous spells of weather and, if so, how.

12.3   Forecasting rainfall

The trouble with trying to provide accurate predictions about the occurrence of rain is that it is typically the product of systems only a few kilometres across. Even in the case of well-defined weather features such as cold fronts, depressions, and tropical storms, there is a great deal of fine structure which can lead to considerable variation in precipitation over quite short distances. The most dramatic variations can occur with showers or thunderstorms, when some places can be drenched by record-breaking rainfall while only a few kilometres away it is dry. An extreme example occurred in London in August 1975 when Hampstead received over 150 mm of rain in two hours while only 10 km away it was dry.

Such small-scale features cannot yet be handled by existing numerical forecasts. The most detailed computer models used for predicting rainfall up to 24 h ahead have a horizontal grid spacing of about 75 km. This means that a great amount of detail will inevitably be smoothed out. Even with expected advances in computer technology it will be many years before there is any prospect of standard numerical forecasts producing accurate predictions of local rainfall. The immediate way forward lies in using a combination of ground-based radar measurements and satellite observations. This technique has already been refined in several parts of the

Fig. 12.1. The 'Great Storm', which was not accurately forecast, did huge damage in southeast England in the early hours of 16 October 1987. This visible image shows the storm at 0819 GMT on the morning of 16 October 1987, a few hours after the strongest winds had crossed the country (with permission of University of Dundee).

148

world to produce accurate local forecasts of rainfall up to a few hours ahead to meet the specific needs of many customers.

In Britain a sophisticated system has been developed which combines data from five weather radars and satellite cloud images from Meteosat 2. It has been operated on an experimental basis for several years and became fully operational in 1986. The core of the system is the weather radars. Working at wavelengths of between 5 and 10 cm they can make accurate measurements of rainfall rate over a distance of approximately 75 km. This performance is the product of over 30 years of development of radar meteorology which has demonstrated that, in spite of practical limitations in making the observations (see Section 8.2), reliable measurements of moderate and heavy rainfall can be obtained from the size of the signals reflected back by the raindrops. Using five radar sites gives an accurate instantaneous picture of rainfall over much of England and Wales.

The images obtained from Meteosat 2 every half-hour give a wider picture of rain-bearing clouds over the British Isles and surrounding areas together with their development and movement over time, but they cannot provide the same accuracy of rainfall measurement as radar. What they can do is fill in the picture through their more extensive coverage and their ability to detect shower clouds before they start to produce heavy precipitation. The forecaster can combine radar and satellite data using a variety of techniques. These include statistical relationships, direct calibration of cloud imagery against radar data, and subjective judgement. The time taken to make these improvements is only about 30 min. The results can be used as very short-range forecasts using either simple extrapolation of the movement of rainfall patterns by the forecaster or by more sophisticated computer analysis. Their value is immediate. For instance, at Wimbledon in 1985 a warning of a sudden downpour enabled groundstaff to take more effective action to protect the courts. Similar benefits have been obtained in such areas as agriculture and flood control.

A similar system has been developed in Japan where almost the entire country has been covered by a network of weather radars since 1971. This has been combined with a mesoscale automatic surface measurement system since 1979 and the operational output of the GMS series of Japanese geostationary satellites since 1978. The aim of this comprehensive system is to provide more accurate forecasts of heavy precipitation associated with tropical storms, like the typhoon Tip (see Figs. 6.3 and 6.4), which often lead to damaging flash floods.

In the United States, observations from the VAS instruments on the GOES satellites have provided additional means of forecasting heavy rainfall. These instruments can detect areas of atmospheric instability where severe storms may develop. Using a combination of visible images and infrared measurements of temperature and moisture it is possible to identify where conditions exist for intense convective activity, enabling forecasters to pick out more accurately areas where storms, and possibly tornadoes, will develop within hours. This approach offers the prospect of extending the use of geostationary satellite data to probe where new developments will occur.

The success in this area suggests that complementary data obtained from radars and geostationary satellites can be used to improve detailed numerical forecasts up to a day or two ahead. This could lead to improved understanding of the factors leading to the development and decay of rainfall systems. But this requires accurate representation of local vertical structure and humidity fields. The best way to do this remains through the use of simple instrument packages which are launched by radio-sonde balloon which transmit data back to Earth as they rise through the

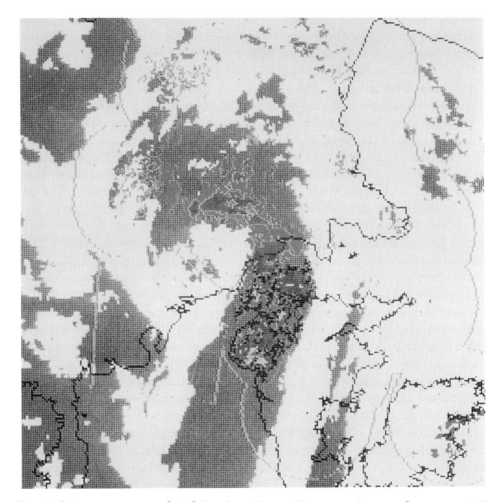

Fig. 12.2. Radar measurements of rainfall can be used to provide accurate short-term forecasts, especially when combined with routine geostationary satellite observations. This example shows a combination of radar and satellite observations over Europe (with permission of UK Meteorological Office).

atmosphere. The accuracy of this standard equipment is beyond the scope of current satellite developments. Measurements of the structure of the atmosphere from space will probably make more important contributions in the extension of the range of predictions, but even here the demands of the forecasting business pose formidable challenges.

12.4   Extending the range of useful forecasts

Forecasters obtained immediate benefits from early satellite images which helped them to interpret and improve their short-range forecasts, but the effective use of quantitative measurements from space took much longer. Two principal sets of observations have been slowly incorporated into the computer models. One of these has been the estimate of wind speed at various levels obtained from the movement of clouds in successive images transmitted from geostationary satellites (see Section 7.6). The other is temperature soundings obtained from infrared and microwave radiometers on polar orbiting, and more recently geostationary satellites (see Section 7.2). In addition, the stream of data from satellite interrogation of automatic instruments in aircraft and on ships and buoys has been of growing value. All of these data sets, which amount to some 8×10^7 bits of information per day, can be used in one way or another in forecasting work. In practice, only a carefully selected set of data can be used in numerical forecasts (see Table 12.2), and their precise contribution in producing better predictions is a matter of conjecture.

A full appreciation of the difficulties in exploiting the potential of satellite observations requires knowledge of how numerical forecasts are prepared. The first step is to produce a computer representation of the atmosphere at some initial point in time. This is normally done twice a day, at noon and midnight GMT, which are termed 'synoptic times'. Since a typical model contains some 15 vertical levels and a horizontal grid spacing of around 100–150 km, this is a major operation. It involves converting all the available data to match the required format. In the case of ground-based observations, which consist principally of some 14 000 sets of surface observations and over 2300 radio-sonde soundings, the coverage is not the same as the horizontal grid spacing, though in the latter, the data are sufficiently detailed to meet the vertical resolution of the model. The majority of the ground-based data is recorded at noon and midnight GMT, so this information can be assimilated directly to provide the initial physical parameters (eg., temperature, pressure, wind speed and direction, and humidity) but the data must be interpolated to provide values at each grid point in the model. Data obtained by satellite interrogation of several thousand drifting buoys, ships, and aircraft can be incorporated in a similar fashion.

A slightly different process applies to satellite sounding data. The continuous, or asynoptic, nature of satellite measurements requires alternative forms of analysis. One approach is to group the data into six-hour blocks and to assimilate them into

Table 12.2. *Sources of data for global weather forecasts produced by the*
UK Meteorological Office in 1986

Platform	Mean number of sets of observations each day
Land stations	14471
Satellites	12208
Aircraft	3265
Ships	2857
Upper air (radio-sondes)	2362
Drifting buoys	1785

In addition later data from such sources as geostationary satellites and
ground-based radars are used by the forecaster to refine the computed
predictions.

the forecast at a single time. The problem with this approach is that appreciable
changes may occur in the real atmosphere between the time when the measurement
is made and the mean time that it is used in the model. The alternative is to incorp-
orate each new observation into the model while it is running. On the face of it
this seems a better approach, but because the new set of readings may be appreciably
different from the corresponding set of values being used in the model one step
earlier this can generate computational shock waves which propagate through the
model and grow into appreciable errors.

There are also snags with the spatial distribution of satellite soundings. In particular, while the geographical spread is excellent, the vertical resolution leaves much
to be desired. For example, in wind speeds inferred from geostationary satellites
there is considerable doubt about the level of the clouds. At best they can be used
to provide wind speeds at three levels (see Section 7.6) and, for weather forecasting,
only two levels (approximately 850–900 and 200–300 mb) are routinely used to provide a measure of movement at the bottom and top of the troposphere. The same
problem occurs with temperature measurements. The altitude ranges of each channel
of the satellite radiometers overlap, and there are only four independent infrared
channels in the troposphere, where the weather occurs, as opposed to about 10 levels
in the computer model. Moreover, in overcast regions, because only the microwave
channels can be used, the number of tropospheric channels is reduced to two. The
resulting error is about 2 °C at best for layers 2–4 km thick (see Table 7.1). Radio-
sondes, for all their drawbacks, measure the temperature continuously through the
10 km of the troposphere with an accuracy of 0.5 °C or better.

Faced with these limitations modelling work in the 1970s could not make full
use of satellite data. The basic problem was that, despite their accuracy, satellite
radiometers produced a rather fuzzy picture compared to ground-based observa-

tions. Where there was good coverage at the surface, notably over land in the middle latitudes of the Northern Hemisphere, the benefits of satellite measurements in computer models were not great. Where ground-level observations were sparse the benefits were much greater. This lack of detail meant that much of the information obtained from space was discarded in favour of less extensive but more precise ground-based results. In recent years improved analytical methods have produced more consistent satellite results. The scale of these benefits has been estimated in a set of experiments by the European Centre for Medium Range Weather Forecasts (ECMWF). Using all available data it has been able to produce useful forecasts out to about seven days in the Northern Hemisphere. This is almost a full day better than forecasts without satellite data – an impressive difference, and space-based data proved indispensable in forecasting at low latitudes and in the Southern Hemisphere.

Two other features of these ECMWF experiments are important. First, the satellite data provided dramatic improvements where major features were forming far from land (eg., in the North Pacific). In some cases these storms could be missed entirely in their early stages by surface-based observations. This meant that the scope of forecasts was inevitably limited by their failure to include the development of such systems. Second, by using all of the satellite data, including the satellite interrogation of surface and airborne instruments, forecast results were improved. Even where observations overlapped and appeared to be redundant they proved to be of value.

A more recent example of the sensitivity of forecasts to having increased information is the 'Great Storm' which hit southern England in October 1987. As noted in Section 12.2 the failure of forecasters to predict its strength only a few hours before it struck led to widespread criticism. Subsequent analysis suggests that if

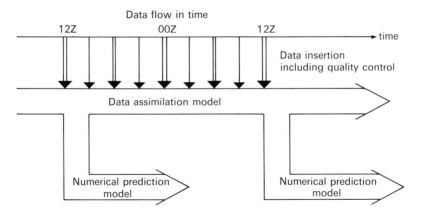

Fig. 12.3. The continuous stream of weather data from many sources needs to be pulled together if it is to be used in numerical forecasting work. This diagram shows how the data is assimilated to provide an input to the computer model at regular intervals.

observations made on commercial airlines over the North Atlantic, which were not received in time, had been incorporated into the forecast a much better job of predicting the explosive development of the storm would have been made. So, on all time scales there is evidence that more complete and timely data is an essential component to improved forecasting.

In the future, satellites will have a central part to play in obtaining more precise and comprehensive observations needed to advance both the accuracy and range of numerical weather forecasts. These advances will include making better use of temperature sounding data. In this context the more subtle inversion technique (see Section 4.3.2) which uses the products of the most recent forecasts to provide an improved 'first guess' of the temperature profile may help. In particular, it circumvents problems which occur when the tropopause is far from climatological normal and so removes gross errors in temperature measurements. Use of this method could produce an accuracy of ± 1 K in satellite-derived temperatures, but it can only go so far as it is dependent on earlier forecasts and so can have little impact on extending the range of the forecast. What it does do, however, is to eliminate grossly inaccurate temperature measurements which might otherwise play havoc with the forecasts.

Even if measurement techniques could be greatly improved, there are limits to what can be achieved. Theoretical studies of current forecasting methods suggest that even with complete information about the initial state of the atmosphere, forecasts beyond about 10–14 days will not be possible. This limit is a consequence of the inherent unpredictability of the behaviour of the atmosphere, where significant local features such as thunderstorms can grow in a matter of minutes out of virtually nothing. Such detail will never be handled in computer models. So, if the way forward does not lie simply in bigger computers using better satellite measurements to produce more precise predictions, is there any hope of extending forecasts to weeks and even months ahead? The answer may lie in obtaining a better understanding of the global climatic system.

12.5   Spells of weather

The prospect of making useful predictions of some broad features of the weather farther into the future depends, in part, on what we call 'useful'. In practice what we require to know is not whether it will be raining in the afternoon of the Wednesday after next, but whether the next few weeks will be out of the ordinary in some way. Because society is finely tuned to the average climate, what causes the greatest problem is a spell of abnormal weather. The ability to anticipate hot or cold spells, drought, or periods of excessive rainfall is therefore of particular value.

The key to progress appears to be in modelling the relatively stable systems embedded in the normal turbulent flow of the atmosphere which are associated with

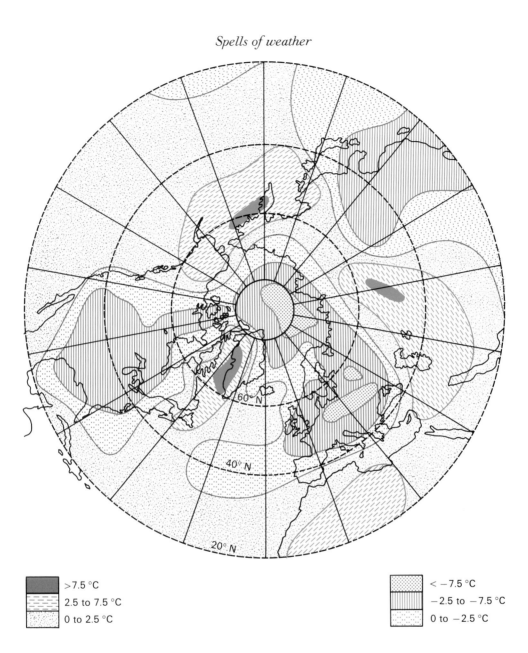

>7.5 °C		< −7.5 °C	
2.5 to 7.5 °C		−2.5 to −7.5 °C	
0 to 2.5 °C		0 to −2.5 °C	

Fig. 12.4. When a well established blocking pattern develops, it can have global implications. The winter of December 1962 to February 1963 is an extreme example of such a global circulation anomaly. This diagram shows that as a consequence of this pattern adjacent regions of the mid-latitudes of the Northern Hemisphere experienced compensating extremes with northern Europe, the United States, and Japan being exceptionally cold and Greenland, central Asia, and Alaska being very mild.

spells of abnormal weather. These features retain their character for considerable periods, so if forecasters could identify what causes them to stay in one place then there would be a good chance of predicting the general features of the weather up to a month ahead. To do this they must understand the cause and behaviour of blocking anticyclones. These enigmatic weather systems, which in Britain lead to extremes such as the hot summers of 1976 and 1983 or the savage winter of 1963 or the cold spells of January 1985 and February 1986, have been the source of frustration to meteorologists for decades.

The properties of blocking anticyclones have been thoroughly studied (see Section 2.2.2). They appear to be essential features of atmospheric circulation in the middle latitudes of the Northern Hemisphere, where they limit the build-up of the westerly winds by splitting the airflow into two streams that run north and south of the block. Most frequently they occur close to the Greenwich meridian and over the eastern Pacific. Atlantic blocks are nearly twice as common as the Pacific variety. Normally they last up to two weeks, but can persist much longer. Once in place, they create a meandering circulation pattern which sets up compensating alternate areas of abnormal temperature and rainfall.

The frustrating feature about blocking is identifying the conditions which cause it. Its preferred position suggests that the distribution of oceans, continents, and mountain ranges in the Northern Hemisphere controls its location. This inference is supported by the fact that in the middle latitudes of the Southern Hemisphere, which is covered mainly by oceans, blocking is less frequent and less pronounced. In this context, the east–west temperature gradient between oceans and continents appears to play an important role. In particular, large-scale sea surface temperature anomalies in the North Atlantic or North Pacific seem to be implicated with the onset of blocking. Other features such as snow cover, surface moisture, and the distribution and extent of arctic pack ice may also be implicated.

Once a block has been established it is, in effect, sustained by the very nature of its circulation. The low pressure systems running around its edge serve to pump up the central high pressure area. This behaviour makes it easier to predict the weather up to a week ahead. But, it does not explain why some blocks subside in less than a week, while others may last well over a month. Again, the disposition of the slowly varying components of the climatic system seems to hold the key to this riddle. Whatever the answer, weather satellites will play an essential part in addressing the problem. Not only do they represent the best way of monitoring the state of the surface conditions which could signal the onset of blocking but they will also enable us to build a better library of the possible array of surface anomalies and the way they are linked to sustained abnormal atmospheric circulation patterns.

Progress in these areas will take a long time. The number of abnormal atmospheric patterns is, in theory, huge. Moreover, links with the different combinations of surface anomalies are complex. A complete library of the alternatives could take hundreds if not thousands of years to assemble! But the fact that, in practice, most blocks occur in certain relatively narrow regions suggests that some clues may emerge much more rapidly. Indeed, there is already evidence that computer models can sometimes identify the critical conditions required for global weather patterns to settle into a quasistationary state. With time, better computer models will draw on improved climatic data to demonstrate increased understanding of how the weather machine functions. This will lead to a demand to improve the data obtained by satellites to make the next generation of models more realistic. These improvements will come not only from improved measurements but also from more accurate representations of the interaction of the atmosphere with the Earth's surface. A good example of how even basic features need careful treatment is in the improvements that have been made in representing mountains in current computer models. On a 150 km grid, mountain ranges such as the Rockies and the Alps average out as 1000–2000 m plateaux and the effect of the higher peaks is lost. In reality, winds high up in the atmosphere have to go round these peaks. An averaging process diminishes the impact of these ranges which play an important part in the position of blocking anticyclones. Including a more accurate representation of the real world by making the mountains higher and filling the valleys with stagnant air has improved forecasts appreciably.

Satellite measurements of year-to-year fluctuations of sea surface temperatures, cloud cover, and arctic ice distribution show the sort of features that must be included in future models. When models can accommodate these fluctuations and demonstrate the ability to reproduce weather patterns associated with certain combinations of underlying conditions, then an interesting possibility could arise. Since

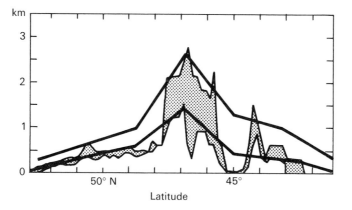

Fig. 12.5. The effect of mountains needs to be carefully handled in computer models. This diagram shows the difference between using the average height of the Alps and making some allowance for the fact that the prevailing winds have to pass over the highest peaks while air is trapped in the valleys between. The latter approach leads to significant improvements in forecasts up to a week ahead.

the record of the sequences of global weather patterns cannot be expanded rapidly, computer simulation may be a better approach. With a big enough computer and a sufficiently sophisticated model of the global climate it may be possible to run an endless series of weather models. Already researchers have run several years of weather simulation in their models and produced realistic patterns of monthly and seasonal weather.

By running literally thousands of years of weather under current global climatic conditions it may be possible to identify guidelines of how monthly and seasonal types follow one another. Armed with these rules meteorologists would then be able to make useful forecasts on the basis of knowing about the current state of the global atmosphere, oceans, and land surface. Thus, in the end, long-range weather forecasting could depend on creating an artificial global climate on a computer which could explore far beyond our limited experience to date. Such an operation will be dependent on improved satellite observations, both to create the model and to operate it. Moreover, there is a major proviso; the model has to assume that the real climate does not change. Here again satellites will play a central role in monitoring any such changes, as we shall see in the next chapter.

13

Climatic change

'Plus ça change, plus c'est la meme chose'
(The more things change, the more they are the same)
Alphonse Karr 1808–90

THUS FAR we have concentrated on how satellite measurements are put to immediate use. Even where observations are of slowly varying components of the climatic system, such as ocean temperatures or the extent of polar ice, the principal interest is in current patterns and how they may influence the weather. While year-to-year variations are important components in the analysis, longer-term trends have yet to attract much attention. This is because the records still only extend over a few years, and the long-term stability of many of the instrumental records has yet to be established with sufficient confidence to define such trends. It is, however, becoming increasingly apparent that satellite data could hold the key to a better understanding of long-term climatic change. This is of interest for two reasons. First, as evidence of past change accumulates, there is reason to believe that an explanation of the changes will be found. With this accumulation of evidence there is the prospect of unravelling the natural causes of climatic variability. Second, mounting concern that the activities of man could lead to a permanent shift in the climate have brought a new urgency to the subject. The impact of the emission of CO_2 from fossil fuels and the damage which chlorofluorocarbons from aerosol sprays and various industrial activities may cause to the ozone layer are examples of man-made pollutants which could change the climate. Satellite monitoring of current climatic trends will be an essential element in studying these related problems.

13.1 Climatic variability

Changes in the climate of the past are of widespread popular interest. Whether it is past Ice Ages or Frost Fairs on the Thames, these climatic aberrations are fascinating and have been the subjects of many books. Here we can only note that on time scales ranging from a few years to the span of geological time the Earth's climate has been changing. This book cannot go into detail, but it can examine those areas where satellite meteorology may make an early contribution.

To examine the areas where satellites may make an impact on our understanding of climatic change we do, however, need to briefly review certain aspects of recent trends. First, there is now clear evidence that the climate in the Northern Hemisphere warmed by about 0.5 °C between the 1880s and the 1940s. It then cooled by about half this amount by around 1970 since when it has been rising sharply. Secondly, global trends in sea surface temperatures have shown broadly the same patterns. Thirdly, there have been significant changes in rainfall over the continents of the Northern Hemisphere since the mid-nineteenth century. Most striking of these are the rise in rainfall in mid-latitudes (35–70° N), notably since 1940, and the sharp decrease in the 5–30° N band since 1960.

Because weather satellites are a recent development their contribution to the study of climatic change will have to fit into existing studies. The first priority of satellite programmes is to investigate changes that occur on a time scale of a few years to a few decades. This work will concentrate on producing better observations of trends in slowly varying components of the climatic system (eg., the global radiation balance, sea surface temperatures, and the extent of snow and pack ice). These observations may hold the key to explaining variations in trends measured by surface instruments. The first results of analysing the observations made by the microwave sounding unit on the NOAA satellites (Section 4.5) between 1979 and 1988 show that it is possible to measure monthly global lower atmospheric temperatures to an accuracy of 0.01 °C. These results confirm the potential of satellites to monitor future global temperature trends.

One area where these observations may play an important role is in exploring whether there is any broad evidence of cyclic behaviour in the climate. Until recently the debate about the causes of apparently periodic climatic change had focussed on the question of whether it was the result of extraterrestrial effects (eg., sunspots, lunar and solar tides, and orbital variations) or simply a product of its own natural variability (autovariance). While the possible impact of human activities has added a new dimension to this debate, the basic argument remains of fundamental interest to climatologists. In defining the scope of the investigation the search for cycles has been essential in establishing the scale of external influences. Sunspots, the tides, and variations in the Earth's orbit all exhibit periodic changes which, if they affect the climate, should be detectable in weather records. In contrast, the effects of climatic autovariance are likely to be more chaotic and show little or no long-term regularity.

The search for cycles has occupied meteorologists for more than a century. Despite a huge amount of work full of tantalising glimpses of what looks like periodic behav-

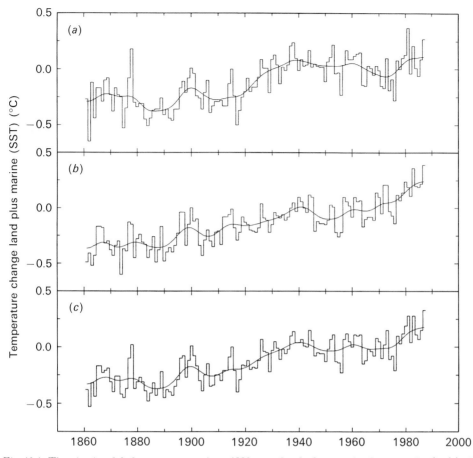

Fig. 13.1. The rise in global temperature since 1880 can clearly be seen in these graphs for (a) *the Northern Hemisphere,* (b) *the Southern Hemisphere, and* (c) *the whole globe (from P. D. Jones* et al., Nature, **322**, 430 (1986)).

iour, there is depressingly little convincing evidence of cycles on the time scale of a few years to 100 years. There is a rather better case for arguing that the Ice Ages are the products of regular changes in the Earth's orbit on the time scale 20 000–100 000 years. There is, however, increasing support for the view that many of the apparently regular fluctuations in the weather may be the product of quasiperiodic interactions between slowly varying components of the climatic system. Monitoring parameters such as cloudiness, sea surface temperature, and the extent of the cryosphere may eventually reveal clues as to how they vary over the years and whether they do so in an ordered way.

13.2 Solar variability

Before considering how the global climate may vary of its own accord, there is one external source of variation about which satellites are making an important contribution. This is in resolving the fundamental question of whether the output of the Sun – the solar constant – is varying. Surface observations over many years have produced intriguing suggestions that the total output of the Sun varied by as much as 1%. But these observations were suspect because it was not possible to remove the effects of atmospheric absorption with any confidence. These objections are removed when satellite measurements are used. While earlier satellites made some preliminary observations of the solar constant, it was not until the launch of Nimbus 7 that it was possible to obtain sufficiently accurate measurements for long enough to make any observations about the climatic significance of solar variability.

The Nimbus 7 results together with observations from a special satellite (the Solar Maximum Mission (SMM)), launched in February 1980 to study changes in solar output following the maximum in sunspot activity in 1980, have shown that the Sun's output does vary by up to a few tenths of a per cent. The biggest changes occur over a week or two. These are too rapid to influence the global climate appreciably, but smaller sustained changes may be more important. By late 1987 it was becoming clear that solar output was in some way related to sunspot number. When

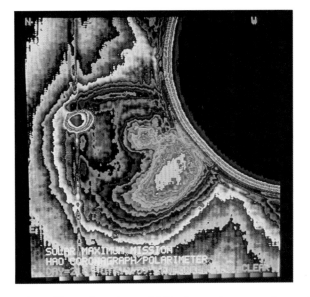

Fig. 13.2. There is growing evidence that solar variability has an impact on the weather. These changes occur on all time scales. Here an example of measurements by a coronagraph–polarimeter aboard the Solar Maximum Mission Satellite on 14 April 1980 shows differences in electron densities in the solar corona over a period of two hours (with permission of NCAR).

the number of sunspots was high in 1980 and 1981, the output was highest. As the sunspot number fell the output declined slightly. Between 1980 and mid-1985 SMM measurements of solar output fell by about 0.08% and then started to rise again. A number of theories have been advanced to explain this behaviour, but until a more lengthy record has been obtained it is premature to assume that there is a direct relationship between sunspot number and solar output. While such changes are not large, over the 11-year period of the sunspot cycle, or over the longer time scale which involves changes in sunspot number from cycle to cycle, they could be climatically significant. If future measurements confirm that there is a simple link, they will hold the key to one of the most elusive questions in climatology – whether there is a real connection between sunspot number and climatic change.

13.3   Global cloudiness

In earlier chapters we have seen examples of observations satellites have made of the slowly varying components of the climate and how they change from year to year. This analysis has shown that more must be done to provide sufficiently reliable series of observations to draw sensible inferences about the nature of climatic change. A good example of the problems of producing a stable set of observations from various satellites is illustrated by the problem of estimating changes in global cloudiness. By 1971 an estimate of the global albedo of 30.0% had been published. This figure relied heavily on estimates of the extent of global cloud cover and the level and depth of the clouds involved. Since then there have been various recalculations ranging from 29.3 to 32.3%. This may seem like a small difference but computer models suggest that a real change of this size could be sufficient to cause climatic changes almost as great as have occurred between the last Ice Age and current climatic conditions.

The uncertainty implicit in these figures demonstrates the problems facing climatologists. The scatter in values, in part reflects the differences in the satellite systems over the years. In particular changes in spatial resolution and optical characteristics of the radiometers used has interfered with attempts to refine the figures. These problems have been further complicated by evidence of instrumental drift. For example, a set of observations of trends in the albedo of the Northern Hemisphere during the 1970s has proven inconsistent with other radiation measurements. This suggests that the performance of radiometers in the visible region can deteriorate over a number of years. Because the visible channels of advanced radiometers were not designed to make absolute radiation measurements, they did not carry on-board calibration systems, whereas the infrared channels did. This meant that small changes in detector sensitivity were not noticed at the time.

These are not the only limitations in the albedo measurements. Analysis of data collected for over 20 years of broad-band radiation measurements show that two other factors need to be considered. First, the fact that many of the albedo estimates have been made using radiation measurements in the visible region (0.5–0.7 μm), results in an overestimation of the albedo of clouds as compared to making measurements over the complete range of solar radiation because clouds scatter more efficiently in the visible region. In addition, over snow-free surfaces change in the planetary albedo due to cloud cover is also overestimated, but over snow-covered surfaces it is severely underestimated. Even under clear sky conditions the visible albedo is a poor approximation to the broad-band figures.

The second limitation occurs in using observations from Sun-synchronous orbits. This introduces a considerable bias, especially in the tropics where many clouds appear during the late afternoon. When combined with the fact that most of the broad-band radiation budget measurements have been made using wide field-of-view radiometers that can miss important features in tropical cloud structure, the overall picture is further confused. Indeed, until recently it was concluded that in spite of having obtained data for 20 years, it was not possible to answer the fundamental question of whether the net radiative effect of clouds on the global climate is one of cooling or heating.

The latest generation of weather satellites has, however, provided the first reliable figures of the overall effect of clouds. This work has combined observations from NOAA 9 plus those from the Earth Radiation Budget Satellite (ERBS). The latter is an experimental system launched from the space shuttle *Challenger* in October 1984 into an inclined orbit which precesses through all local hours at the equator in 36 days. This combination has reduced temporal sampling errors. The first results obtained from analysis of data for April 1985 show that the net global effect of clouds is to cool the climate. The size of this cooling (13.1 watts (W) per square metre) is considerable given that it is estimated that the net radiative effect of a doubling of CO_2 in the atmosphere is about 4 W m^{-2} which could lead to an increase in the global surface temperature of between 1.5 and 4.0 °C. So any changes in cloudiness could play as important a part in future climatic changes as the build-up of carbon dioxide in the atmosphere.

Preliminary results for July 1985, October 1985, and January 1986 produce similar results, with if anything the net cooling effect being greater in these months and highest in winter. This suggestion of seasonality underlines the potential complexity of any possible future changes in cloudiness. Moreover, the results for April 1985 show that the net effect of clouds varies with latitude. In the tropics the impact is small. In mid-latitudes there is a marked cooling, whereas in the polar regions there is a slight warming. So the impact of any changes in cloudiness may also depend on shifts in global weather patterns. In this context, these recent measurements have confirmed some of the results obtained from earlier observations. Using measurements of both reflected sunlight and outgoing terrestrial radiation covering the period from 1964 to 1977 it was possible to produce an interesting view of the way the radiation balance of the Earth varied from place to place and with the seasons.

The overall annual figures confirmed the standard model of net income of solar energy at low latitudes being balanced by outgoing long-wave radiation at high latitudes (Section 2.1.4). Calculation of these annual averages did, however, provide an indication of the complexity of the processes involved. In particular they provided the first global figures on the balance between how much incoming solar energy clouds reflected as opposed to how much terrestrial radiation they trapped. Climatologists have long puzzled over which of these effects dominates in different circumstances and the role of changing cloud cover and type of seasonal and inter-annual fluctuations in the weather. For instance, the cold tops of towering clouds associated with intense convection in the tropics radiates little energy and this compensates their high albedo. The observed balance of these effects may be vital to maintaining the Earth's climatic equilibrium. But for low flat clouds which are relatively warm radiators and effective reflectors, the net influence is definitely to reduce the amount of energy at the surface.

Between latitudes of 30° N and 30° S, the observations show that even annual patterns of radiation balance are complex, and show the influence of different physical processes. The most pronounced difference is between the oceans and the continents. One striking feature is the high net radiation gain over southeast Asia associated with the intense convection of the summer monsoon. By contrast the low values west of South America and South Africa are due to persistent layers of low cloud which form over the cold ocean currents and reduce the net warming of the tropics. In this band of latitude it is the subtropical deserts which gain least radiation. They have highly reflective surfaces, which also get hot enough to be very effective radiators, so radiative losses tend to cancel out solar heating gains. In the extreme case, the Sahara desert – archetypal example of a hot place – actually experiences a net loss of energy from the Earth over the whole year. During the day it reflects a lot of sunlight, while at night, with virtually no cloud cover, the ground radiates its stored heat rapidly. In the rare instances when clouds form in this region their net effect is a warming. Moving toward the poles, the pattern becomes much simpler, because outside the central latitude band solar input can never match outgoing thermal radiation. Indeed, observations confirm the basic analysis in Chapter 2 about net losses at high latitudes being matched by energy transfer in the oceans and the atmosphere from the tropics, but the process is far from uniform and predictable.

The same measurements have shown significant seasonal variations in cloudiness, notably in the tropics. These are greatest in the Asian monsoon region, the North Pacific, west of South America, and over Amazonia and the West Indies. In part this is due to seasonal movements of the Intertropical Convergence Zone (Sections 2.2.1 and 6.5). The variance of the extent of cloud cover which forms over cold

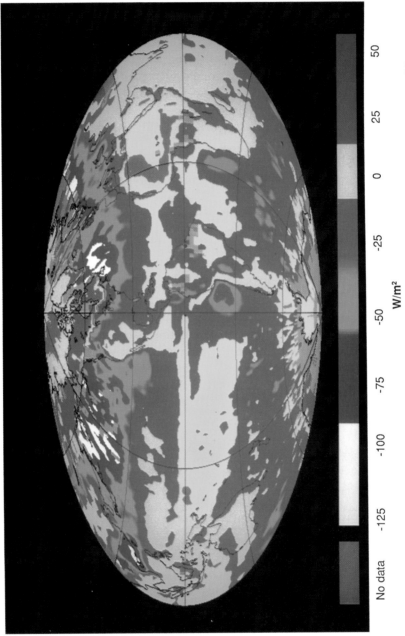

W/m²

No data   -125   -100   -75   -50   -25   0   25   50

Fig. 13.3. The net effect of clouds on the radiation balance of the Earth is complicated. Measurements made from satellite during April 1985 show that in the tropics the amount of sunlight reflected back into space by clouds is roughly balanced by the terrestrial radiation they trap. In mid-latitudes the reflection effects dominate especially over the North Pacific and North Atlantic, whereas over snow-covered polar regions clouds have a warming effect (with permission of the University of Chicago).

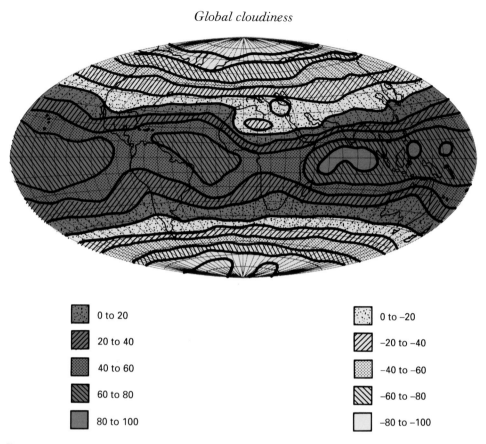

0 to 20	0 to –20
20 to 40	–20 to –40
40 to 60	–40 to –60
60 to 80	–60 to –80
80 to 100	–80 to –100

Fig. 13.4. Satellite measurements over the years show how the net balance between incoming solar radiation and outgoing terrestrial radiation varies with lattitude and longitude. While these observations show considerable variations between different parts of the tropics and subtropics, the important feature is that energy is pumped into the climate system at low latitudes and escapes at high latitudes.

ocean currents is a key factor in determining the weather patterns over the eastern Pacific. The third cause of these variations is associated with shifts between the wet and dry seasons in subtropical regions. These observations, together with more recent results, underline the importance of knowing more about global cloudiness. Some surface-based observations in the Northern Hemisphere suggest that the warming trend during the first half of this century was mirrored by an increase in cloud in the mid-latitudes. Given that the net effect of these clouds is to cool, an increase could counteract the warming effect of the build up of CO_2 in the atmosphere. Thus improved measurements of cloud cover are central to forecasts of the overall impact of current changes on the future global climate.

One further aspect of the potential natural variability of the climate has been opened up by satellite observations. This is the impact of large volcanic eruptions which can form an extensive veil of dust and aerosols in the stratosphere lasting for several years. Climatologists have argued for many years that major eruptions have caused periods of below average global temperatures. In 1982 the eruption of El Chichon in Mexico produced the most extensive dust veil recorded in 20 years. Observations of the dust cloud were made by the stratospheric sensors in Nimbus 7 and measurements of its impact on sea surface temperatures were made by the AVHRR. These showed that the cloud had an appreciable effect on the radiative properties of the stratosphere, but these observations, together with surface measurements, did not provide unequivocal evidence of the impact of volcanic eruptions on the climate. We will have to wait until a really big eruption resolves this particular climatic debate.

Apart from detecting natural variations there is another point in making satellite measurements of cloudiness. This is to explore the effect of man-made particulates on the properties of clouds. At the simplest level it would be expected that since most of these pollutants are dirty or sooty they will absorb more sunlight, so the obvious result would be an overall warming of the climate. But things are not this simple. The reflectivity of clouds depends on both the size of the droplets and their number. More small droplets are more reflective than the same amount of liquid water in the form of fewer larger droplets. So if pollutants increase the number of droplets in clouds then the net effect could shift the effect from net additional absorption to increased reflectivity.

Over land it is difficult to detect the effect of man-made pollutants because of the large number of different sources. Over the sea, however, effects of single sources can be observed by detecting the impact of ships. Satellite images have shown that under stable meteorological conditions the effect of ship exhausts on the overlying clouds is to enhance cloud reflectivity in the infrared (as observed by the infrared channel (3.7 μm) of the NOAA–AVHRR). In the visible region (0.63 μm) the reflectivity for contaminated clouds is also significantly higher despite the likelihood that exhaust is a source of particles which absorb at visible wavelengths. The conclusion drawn from these observations is that the effect of man-made particulates on cloud reflectivity is likely to have a larger influence on the Earth's albedo than on the absorption of sunlight. This is an important result which needs further examination as it has major implications for our understanding of the impact of human activities on the global climate.

13.4 Sea surface temperature records

By virtue of their thermal inertia and the long-term nature of their motions, the oceans play a central role in climatic change. Thus predictive models must be able to integrate the behaviour of the oceans with that of the atmosphere. This is not yet possible in anything but the most basic way. Moreover, we have obtained only a fragmentary view of the way in which the oceans have changed in the last 100 years or so, and this limited picture applies only to the surface. What has been happening below the surface is still completely unknown. In Chapter 10 we examined how satellites have started to make useful measurements of sea surface temperatures, and the major currents. These are slowly being built into a substantial data base, but it will be an appreciable time before they can provide useful predictions of long-term trends. This delay is due in part to limitations of satellite observations (see Section 10.1) and in part to difficulties in establishing a reliable surface record. Efforts to standardise the millions of SST measurements made since the mid-nineteenth century have highlighted the problems inherent in the comparability of different measurements. Only with a great deal of detective work has it been possible to produce series which give reliable pictures of global and hemispheric trends over the last century or so.

Only when there is a substantial overlap between surface data and satellite observations will it be possible to use satellite measurements as the basic data for analysing global trends. Nevertheless, recently published results for the period of six and a half years from early 1982 to mid-1988 have demonstrated the potential of satellite measurements. Using data from the AVHRRs on the NOAA weather satellites and calibrating these against drifting buoy data, monthly figures have been produced for the oceans between 60° N and 60° S. The overall conclusion is that there has been a gradual but significant warming trend of about 0.1 °C per year. More important, this is roughly double the trend observed using conventional ship and buoy data. But these results have been the subject of controversy. Because they coincided with the El Nino in 1982/83 (see Section 10.3), and more important, the eruption of El Chichon (see Section 13.3), there is doubt about the accuracy of the early measurements and the observed trend.

Another feature of these observations is the measurement of trends for different oceans. These show the North Pacific and the North Atlantic warming more rapidly than their southern counterparts while the Indian Ocean shows no significant trend. This regional analysis is of value since the differential behaviour of the oceans may play an important part in global weather anomalies on a time scale of months to years. For instance, computer models of the global climate developed by the UK Meteorological Office suggest that the drought in the Sahel between the late 1960s and early 1980s is linked to warming of the southern oceans, notably the South Atlantic, relative to the oceans of the Northern Hemisphere. The mechanism at work appears to be that these changes have influenced the position of the Intertropical Convergence Zone.

13.5   Changes in the cryosphere

The gradual compilation of a lengthy record of satellite observations of the extent of snow and sea ice in the Northern Hemisphere and sea ice in the Southern Hemisphere has been discussed in Chapter 11. Clearly this record is of growing importance in understanding year-to-year fluctuations in the weather. It is also a sensitive indicator of more sustained changes in the climate. It is widely argued that together with SST observations, these changes may be the most accurate indicators of the anticipated global warming due to the build-up of carbon dioxide in the atmosphere. Computer models predict that the warming will be greatest in the Arctic. One of the first consequences of such warming would be a decline in the extent of continental snow cover. Thus far, there has been little clear evidence of such a change. While the extent of both arctic and antarctic ice cover seems to have declined a little during the 1970s, satellite measurements during the 1980s up to February 1989 showed no marked trend one way or the other.

The related but longer-term variation of the volume of the ice-caps of Greenland and Antarctica is even less clear. It is, however, argued that the global rise in the sea level of 5 cm since 1940 is, in part, a consequence of the melting of these ice-caps. The complete melting of all this ice could lead to a rise of about 70 m in the sea level, resulting in the inundation of many of the world's major cities, so measurements like those described in Section 11.6 will be of immense importance.

Fig. 13.5. A processed digital image of the Augustine volcanic eruption in Alaska in July 1986, recorded by NOAA 9. Ratios of the different spectral channels have been used to detect the different properties of the volcanic plume (orange) from atmospheric clouds (green) (with permission of Michigan Technological University).

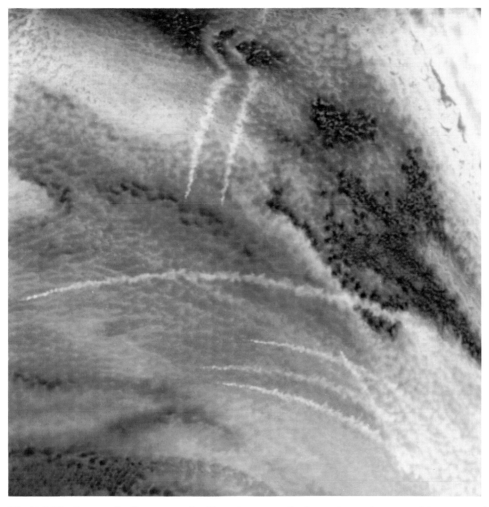

Fig. 13.6. The impact of pollutants on cloudiness is most easily detected over the oceans. The impact of smoke from ships in the Pacific off the coast of California can clearly be seen by the enhanced cloudiness in this image recorded by the 3·7 μm channel of the NOAA 9 AVHRR on 3 April 1985 (with permission of NCAR).

13.6 Changes in land use

Various aspects of the climatic consequences of shifts in land use were reviewed in Chapter 9, but the overall effects of these changes were not analysed. There is evidence that past man-made changes in the vegetative cover of many parts of the world may already have caused a significant perturbation of the global climate. Moreover, the increasing exploitation of natural resources, including pressures to farm marginal areas of land, are problems accelerating this process. The most important aspect of surface changes is the long-term impact of agriculture and the associated clearance of forests. Over the past 6000 years man has cleared a large proportion of the forests in the mid-latitudes of the Northern Hemisphere and has also over-grazed the northern subtropics of the Old World. From the time of Plato there have been frequent records of forest removal. These modifications to the Earth's surface have led to a change in its overall albedo, and in regions where winter snow is prevalent, and to changes in seasonal variation patterns – snow-covered forest has a much lower albedo than snow-covered fields (see Table 13.1). Estimates suggest that about 5% of the Earth's surface – 17% of the continents – has been modified.

These anthropogenic changes could significantly affect the overall radiation budget and modify regional climate. After making suitable allowances for cloud cover, their cumulative effect on the surface albedo is estimated to have increased the global figure by roughly 0.5%, from approximately 0.305 to 0.31. Climatic modelling exercises suggest that this could be sufficient to produce a global cooling of as much as 1 K, which might be enough to explain much of the reduction in temperature that has taken place over the past 6000 years since the post-glacial climatic optimum. Because these effects have accelerated in the last century they need to be considered alongside the much more widely publicised man-made climatic impact of carbon dioxide. In so doing improved satellite measurements of the Earth's albedo will play a crucial role.

13.7   Holes in the ozone layer

Since the early 1970s there has been animated debate about the effect of man's activities on the ozone layer. At various times the chemical arguments have pointed the finger at the exhausts of high-flying aircraft, the increased use of fertilisers, and the use of chlorofluorocarbons as aerosol propellants, industrial cleaning fluids, and in refrigerators, as being potentially damaging. In recent years the debate has focussed on the role of chlorofluorocarbons and their destruction of ozone over Antarctica.

Satellite-borne measurements have played a major role in identifying how the ozone levels in the stratosphere over Antarctica change each October. The fact that these changes occur at the same time each year is central to the debate, because it shows that they are part of a complex chemical process involving the effect of sunlight on various molecular species that have accumulated during the antarctic winter. The return of the Sun at the end of this dark period leads to chemical reactions which result in the sudden depletion of ozone in the stratosphere. There has been a dramatic decline in ozone levels each October during the 1980s. The vertical and spatial distribution of ozone has been measured continually since 1979, using the Solar Backscatter Ultraviolet Spectrometer (SBUV) and Total Ozone Monitoring System (TOMS) on Nimbus 7 (see Section 4.3.6). These observations show

Table 13.1. *How man changes the albedo*

Process	Land type change	Change in albedo	Percentage of Earth's surface affected
Desertification	Savanna to desert	0.16–0.35	1.8
Temperate deforestation	Forest to field/grassland		1.6
	Summer	0.12–0.15	
	Winter	0.25–0.60	
Tropical deforestation	Forest to field/savanna	0.07–0.16	1.4
Urbanisation	Field/forest to city	0.17–0.15	0.2
Salinisation	Open field to salt flat	0.10–0.25 to 0.5	0.1

that the October values of total ozone in the atmosphere over parts of Antarctica declined by over 50% between 1979 and 1987.

In addition, satellite observations have shown temperature drops as large as 18 °C associated with regions of the greatest ozone loss, a change roughly consistent with what might be expected on the basis of the infrared radiative properties of ozone. This figure is, however, based only on data for the period 1979–86. On the other hand, detailed analysis of radio-sonde data for the period 1958–86 shows smaller but significant temperature decreases of 6–8 °C. Balloon-borne studies of the structure of the ozone level over Antarctica have also provided information about the extraordinary changes taking place in recent years. In both October 1987 and October 1989 ozone levels at altitudes of around 15–20 km had been depleted by more than 90% compared to normal concentrations.

These observations provide evidence of a more complicated phenomenon. Scientists have concluded that this decline in ozone levels is largely caused by chlorofluorocarbons (CFCs). The explanation of these changes depends on how ozone is created and destroyed in the atmosphere. It is formed by the action of ultraviolet solar radiation which breaks oxygen molecules into two oxygen atoms. These can combine with oxygen molecules to form ozone molecules which contain three oxygen atoms. But ozone is itself reactive and in the presence of certain other chemicals in the atmosphere it can combine with them and revert to oxygen. In normal circumstances this chemical cycle produces a maximum concentration of ozone in the stratosphere at altitudes of around 15–20 km. The amount present varies with the time of year and global atmospheric circulation, but until the appearance of the ozone hole these fluctuations remained within fairly well-established limits.

How CFCs can interfere with these normal processes depends on a complicated set of chemical reactions. CFCs break down in the upper atmosphere to form highly reactive chlorine compounds. Their global atmospheric concentration has more than doubled in the last 10 years and they have produced a four-fold increase in chlorine compounds compared with pre-industrial levels. These changes are, however, not enough to explain what is happening over Antarctica. According to standard chemical theory the build-up of CFCs might be expected to lead to a global reduction of ozone of around 1%. To account for the observed decrease in ozone levels new theories have evolved as improved measurements have been collected. These theories centre on the peculiar conditions that develop in the antarctic winter vortex. The intense cold, often below $-90\,^{\circ}$C at altitudes around 15 km, produces clouds of ice crystals which accelerate the depletion of ozone through a complex process of surface chemistry.

Major international scientific programmes of ground-based and airborne experiments in 1986 and 1987 have now obtained detailed measurements of the important chemical species involved. These confirm that there is a dramatic difference between the chemistry of air trapped in the polar vortex and that in regions outside it, suggest-

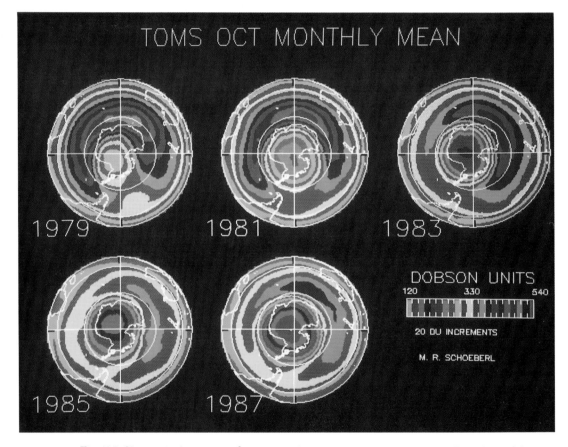

Fig. 13.7. Changes in the amount of ozone over Antarctica were monitored by the Total Ozone Mapping Spectrometer (TOMS) on Nimbus 7 between 1979 and 1987. These images show the measurements for October in 1979, 1981, 1983, 1985, and 1987 and provide clear evidence of how the 'ozone hole' has developed. The hole is the oval feature generally covering the Antarctic portrayed in grey or violet colours (with permission of NASA).

ing that to a large extent the effects of CFCs on ozone will be confined to Antarctica. However, in October 1988 the hole was much less pronounced. This indicates that current chemical models will need to be refined to take account of other factors. In particular, the larger-scale dynamics of the stratosphere, which affect how much the air over Antarctica is isolated in winter from the rest of the global atmosphere, need to be incorporated. In addition, it may be necessary to include a correction for changes in solar activity as there is evidence that the 11-year cycle in this activity affects the rate of ozone production in the upper atmosphere.

There is further evidence of the complexity of the effects leading to the ozone hole. Observations over the Arctic during the winters of 1987–88 and 1988–89 appear to support the basic chemical model. Here the polar vortex is less intense and prolonged, so the temperature does not fall to antarctic levels and the formation of stratospheric ice clouds is much less common. This means that chemical conditions do not become suitable for the rapid destruction of ozone. None the less, both aircraft and satellite observations in 1988 and 1989 do hint at the first signs of a similar but less extreme depletion of ozone in the Arctic at the end of winter.

Clearly, the sudden development of the ozone hole and the rapidly evolving scientific explanations of its origin mean that our understanding of the disturbing phenomenon will continue to improve with more observations. In this context, satellite measurements will have a central role, as in the past. In particular, the fact that the larger-scale dynamics of the global stratosphere play in the formation and size of the hole means that any analysis must be based on a complete picture of how ozone levels change over space and time. Only satellite observations can provide the right amount of information to unravel the puzzling dynamical behaviour of the ozone hole.

As a final comment on the role of satellites in the 'ozone hole saga', it is interesting to note that the original discovery of the hole was the product of ground-based work by Joe Farman and colleagues from the British Antarctic Survey. When the evidence of ozone depletion over Antarctica was published in *Nature* in 1985, the obvious question was why had it not already been seen in the Nimbus 7 observations. It subsequently emerged that the NASA computer had been programmed to reject any ozone figures that were considered to be outside the bounds of physical probability, so the vital satellite data were stored away as being 'erroneous', whereas in practice the development of the hole had been accurately measured since 1979. This oversight highlights two problems of integrating satellite data into studies of climatic trends. First, it shows that unless there is good corroborating evidence from surface observations it is likely that unexpected observations may be discarded on the assumption that they are the product of instrumental malfunction. Secondly, there is always the danger that in the flood of data from weather satellites some potentially important information may be overlooked because of the way in which computers are programmed to select only what is seen as essential information.

14

The future

'I have seen the future, and it works'
Lincoln Steffens 1866–1936

THE RAPID DEVELOPMENT of satellite meteorology in the past 25 years provides a guide to future developments. The technologies described in earlier chapters will be refined and extended to provide improved observations of the world's weather. These advances will involve greater precision, resolution, and frequency measurements. In addition, new techniques currently being developed in the laboratory will be introduced into use. This progress will run in parallel with more sophisticated data handling and analytical methods which will make thorough use of the observations. One way to review what these developments will involve is to consider the new satellite systems that are at various stages of development, but before doing so, it is important to consider the background against which these advances will take place.

The planned development of more advanced satellite systems is clouded by two important factors. First the *Challenger* disaster in January 1986 has resulted in a major delay in large areas of the NASA programme. As a consequence many experimental proposals may be delayed by several years. This delay is the product of not only the loss of momentum in the Shuttle Programme but also a shift in favour of military work. The second drawback is more basic still – the escalating cost of many of the planned programmes. As the planned systems become more sophisticated, the scope for delays and cost overruns grows almost exponentially. The cancellation of the US Navy Remote Ocean Sensing System (NROSS) in January 1987 is a good example of this problem, and affects NASA plans for the 1990s. The effect of delays and financial pressures will alter the pace and direction of developments but will not bring them to a halt. What is more, other approaches may provide partial compensation for these setbacks. In the United States increasing pressure to involve private sector finance in developing commercially viable systems may lead to jointly funded projects and greater emphasis on the application of results. Elsewhere the growing efforts, notably of European and Japanese programmes, could serve as important elements in the development of satellite meteorology in the 1990s. Whatever the combination of effort, the indisputable fact is that this area of space technology is bound to take major strides forward on a variety of fronts in the next decade.

14.1   Refinement of operational systems

The continuing operational systems (TIROS-N/NOAA, GOES, and the other geo-stationary satellites) will be maintained with the launch of a series of new satellites at regular intervals (eg., GOES 7 was launched in March 1987). For the most part these will use the equipment of earlier satellites, but inevitably refinements will be made as reliable improvements are achieved. The NOAA programme includes plans for approximately biennial launches until well into the 1990s. Later spacecraft in the series from 1992 onward will include technical improvements. The AVHRR will be expanded to include a sixth channel in the 1.5–1.78 μm region which will be used to discriminate between clouds and snow cover. This will share time with the 3.7 μm channel (see Table 4.1) making observations of reflected sunlight by day, while longer wavelengths will be observed at night. At the same time the spectral response of the visible channels will be sharpened and the signal-to-noise ratio improved three-fold. This will lead to a variety of improvements including better vegetation index calculations.

Another major improvement will be to replace microwave and stratospheric temperature sounding units with a more sensitive 20-channel microwave system. This will have two low frequency channels (18.7 and 31.4 GHz) to measure surface temperature and one at 23.8 GHz to measure water vapour. There will be 12 channels in the 50–58 GHz region to sense atmospheric temperatures from the surface to high in the stratosphere. The remaining channels at higher frequencies, including three in the 183 GHz line, are principally designed to measure water vapour.

Beyond 1991 the GOES satellites will be uprated considerably. The imaging and temperature sounding systems will be independent of one another. There will be a general expansion of the number of channels with the imager having five and the temperature sounder being increased from 12 to 14. Spatial resolution will be improved from 8 to 4 km in the infrared imager and from 14 to 7 km in the temperature sounder. In addition, the equipment will be able to locate the precise position of individual small-scale features with greater accuracy, improving from the current figure of 5 km to 2 km. These advances, combined with more flexible operating arrangements and more sensitive detectors, will improve the monitoring of atmospheric moisture and stability and lead, in particular, to more accurate forecasting of severe storms.

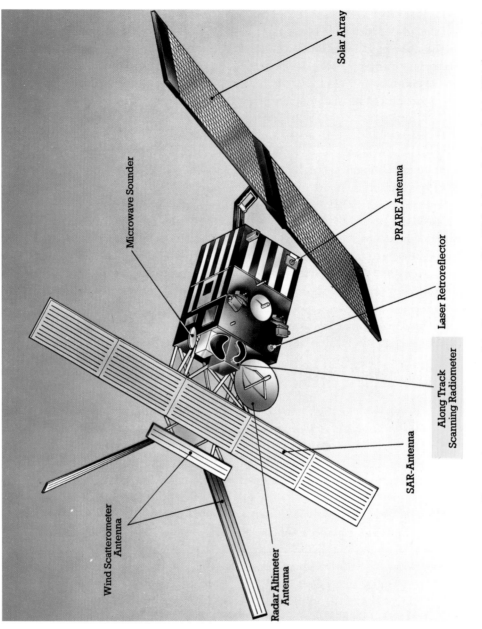

Fig. 14.1. Schematic diagram of the European satellite ERS 1 due to be launched in 1991 (permission Rutherford Appleton Laboratory).

14.2   New programmes for the 1990s

The clearest picture of the potential progress of satellite meteorology can be gained by reviewing the current status and content of the various programmes to be launched in the 1990s. Not only do these show what can realistically be expected from new equipment, but they also provide a good idea of the issues that are of central interest to meteorologists. The most ambitious programme is that proposed by NASA in collaboration with various other US agencies and other countries. But European and Japanese plans will play an increasingly important part in the expanding efforts to understand and monitor the global climate from space.

With the demise of NROSS, the new flagship of the NASA plans is the Ocean Topographic Experiment (TOPEX) which is merged with the French Centre Nationale d'Etudes Spatiales (CNES) Poseidon Experiment. Planned for a 1991 launch, the goal of the mission is to increase substantially the understanding of global ocean dynamics by making precise sea-level observations for at least three years. The mission will use two altimeters. The NASA instrument will be similar to those flown on Skylab, GOES 3, SEASAT, and GEOSAT, except that it will operate at two frequencies so that corrections can be made for errors caused by fluctuations in the ionosphere. The CNES altimeter will be the first of a new solid-state design which will be tested for use on future satellites. The satellite will also carry a three-channel microwave radiometer to gather data necessary for correcting the altimeters' height measurements for the influence of variations of water vapour in the troposphere. It is hoped that together with improved satellite tracking arrangements it will be possible to measure the ocean surface to an accuracy of ± 2 cm, nearly two times better than the accuracy currently achieved with GEOSAT.

Although NROSS has been cancelled, a great deal of work has already been done on the principal experiment for the satellite – the NASA Scatterometer (NSCAT) – which may well be flown on another spacecraft. This instrument, which is an advanced version of the SEASAT system, with six fan-beamed antennas, will allow radar back-scatter measurements at three different azimuth angles within two 600 km wide swathes separated by a subsatellite gap of 350 km. It will have a spatial resolution of 25 km and yield wind speeds and direction with accuracies of ± 2 m s^{-1} and $\pm 20°$, respectively, over the range 3–30 m s^{-1}, and will be able to cover 90% of the ice-free oceans every two days.

An important component of future European efforts is the European Research Satellite 1 (ERS 1). It is seen as the forerunner of a series of European remote sensing satellites operating from the mid-1990s onward. The major feature of the satellite will be a set of radar instruments designed to observe the surface wind and wave structure over the oceans. These will consist of a wind scatterometer designed to measure wind speed and direction; an altimeter to measure significant wave height and wind speed below the satellite and provide measurements of ice and major ocean currents; and a synthetic aperture radar system to take high resolution all-weather images over the polar ice-caps, coastal zones, and land areas. The latter will also operate in a sampling mode over the oceans as a wave scatterometer with the aim of measuring the wavelength spectrum of the ocean waves.

In addition to the radar instruments the equipment will include: laser retroflectors for accurate tracking of the satellite and radar altimeter calibration; an along-track scanning radiometer with three infrared channels (3.7, 11, and 12 μm) and two-channel microwave sounder (23.8 and 36.5 GHz) to provide information about the amount of water vapour in the atmosphere and to correct the radar altimeter observations. The benefit of the along-track radiometer is that it will make measurements of radiance of the same part of the sea surface at different angles to the nadir. These will include different atmospheric path lengths which will allow better correction for atmospheric absorption, especially in tropical regions, and so increase the accuracy of sea surface temperature determination. The expected performance of these systems is given in Table 14.1.

On the research front there are no immediate plans to fill the gap left by the end of the highly successful Nimbus programme. Hopes are now centred on the ambitious Earth Observation System (EOS) in the United States. The first satellite is planned for launch in 1996 and is designed to be a giant step forward. Its payload will include novel electro-optical frequency shifters to obtain middle atmosphere wind speeds. Its temperature sounding system will make rapid three-dimensional scanning of the atmosphere to reveal detailed information about its wave and eddy structure. Also high-resolution cryogenic interferometer spectrometers (see Section 4.4) will be able to measure trace gases in the troposphere and stratosphere.

14.3   New instrumentation

The satellite equipment of the 1990s will largely use instruments which have already been developed. For example, the radar equipment for TOPEX/Poseidon and ERS 1 is only planned to be a limited advance on the experimental devices which were flown on SEASAT, but beyond these developments there is a huge range of more sophisticated systems which are still at the laboratory stage or in the conceptual design phase. In many cases these advances will consist of more complicated versions of the existing instrumentation. More sensitive detectors and improved data-handling techniques will enable equipment with a large number of spectral channels to be developed. This offers the prospect of improving, say, the sensitivity of atmospheric temperature measurements and increasing vertical and horizontal resolution. But, such advances cannot overcome the fundamental limitations of the existing measurement techniques. So the objectives of considerable improvements in the measurement of atmospheric temperatures, water vapour content, and wind speeds, will not be met simply by improving existing techniques.

Table 14.1. *Performance of ERS 1 satellite systems*

Parameter	Range	Accuracy	Instrument
Wind speed	4–24 m s^{-1}	± 2 m s^{-1} or $\pm 10\%$[a]	Scatterometer and
Wind direction	360°	$\pm 20°$	altimeter
Wave height	1–20 m	± 0.5 m or $\pm 10\%$[a]	Altimeter
Wave direction	360°	$\pm 15°$	Synthetic aperture
Wavelength	50–1000 m	$\pm 20\%$	radar
Sea surface temperature	500 km swathe	± 0.5 K	Radiometer
Atmospheric water vapour	25 km spot	$\pm 10\%$	Microwave sounder

[a] Whichever is the greater.

One potentially exciting new development is to use lasers to sense atmospheric properties. In particular, there are proposals to develop a laser doppler anemometer to provide accurate direct measurements of wind speed. The apparatus will measure the frequency shift of laser light reflected from aerosols, clouds, or dust in the atmosphere. By measuring the time taken for the scattered light to return to the satellite and the frequency shift, it will be possible to locate the height of the scattering particles and the speed at which they are moving. The problem with this technique is that, in probing the clear atmosphere, so little light is reflected back to the satellite from any given layer that a very powerful pulse laser is needed. This means the apparatus will be large, require a lot of electrical energy, and be very expensive. Plans to build such a device in the 1970s to fly on the Space Shuttle were scrapped for this reason, but work continues in the hope of developing more efficient lasers and in the knowledge that this technique may be the only remote sensing method capable of providing meteorologists with accurate measurements of global wind speeds throughout the atmosphere.

Lasers may also offer the possibility of improved altimetry, especially over the ice-caps of Greenland and Antarctica. Because microwave instruments cover a large area it is difficult to get accurate ice surface elevations, especially over mountainous regions. A laser altimeter would have a smaller footprint and could more readily detect thickening or thinning of the ice sheets. This could be of particular value over Antarctica where so far radar systems have been unable to detect any changes (see Section 11.6).

14.4 A long look backwards

The future development of satellite meteorology will not simply be a matter of forging ahead with new programmes. The work that has already been done has produced a huge storehouse of data. Much of it has not been fully analysed. In planning future systems and allocating resources, it is important that meteorologists make full use of the data they already have. This will involve continued work on the archives of both satellite observations and parallel surface data.

The scale of the work can be gauged by the fact that SEASAT data are still being analysed. The fact that three months of observations from this advanced satellite kept scientists busy for 10 years shows that major efforts will have to be directed to ensuring improved data analysis methods are available to handle the much greater flow of observations in the future. This demand is compounded when the satellites operate for longer periods. The stunning success of Nimbus 7 is a case in point. It has operated for six years longer than planned and provided vastly more data than expected. In particular, the observations of polar ice and stratospheric ozone have been used to monitor important longer-term changes in the global climatic system, in a way that was not planned as part of the original experiment. But, if this bonus is not properly exploited there is a risk that future satellites will be used in a way which does not take full advantage of their potential.

14.5   Global climate programmes

The way in which the meteorological community organises the interpretation of satellite data will be a decisive factor in the successful application of satellite observations to real human problems. In this context the establishment of worldwide programmes, which pool the talents and efforts of all countries is an important element. Central to this will be the World Climate Research Programme (WCRP) being organised by the World Meteorological Organisation (WMO), which is the successor of the Global Atmospheric Research Programme (GARP) of the 1970s. The planning of the WCRP is already well under way and consists of three principal components. The first has the aim of understanding the physical basis of long-range weather forecasting out to a month or two ahead. The second is to understand and model variations in global atmospheric circulation and their connection with the broad transient changes in circulation modes of the tropical oceans with the aim of predicting interannual changes in the weather. The third aims to understand decadal and longer climate trends.

The first area of effort will concentrate on the role of sea surface temperature and the extent of polar ice in defining the weather a month or two ahead. The major requirement will be the improvement of SST measurement techniques from space. The second phase will be built round a programme to study the interannual variability of the Tropical Oceans and Global Atmosphere (TOGA). The objectives of this work are to determine the extent to which the behaviour of the tropical oceans to global atmosphere is predictable on time scales of months to years; to study the feasibility of modelling such behaviour; and, if feasible, to design an operational forecasting system. TOGA began in 1985 and is designed to run for 10 years in order to span one, and if possible, two cycles of anomalous tropical circulation such as the 1982–83 El Nino. The first of these events occurred during 1987.

The TOGA observation programme will involve 10 years of measurements of the month-to-month variability of temperature, circulation, and pressure fields of the upper layer of the tropical oceans between latitudes of around 20° N and 20° S. In parallel a similar set of measurements of the global atmospheric circulation, thermodynamics, and hydrological cycle will be taken. Linking these will be studies aimed at measuring the fluxes of momentum, heat, and moisture across the air–sea interface. These measurements will rely on the new generation of oceanographic satellites together with improved ground-based systems. They will be used both as a basis for the modelling work and to evaluate the success of such work.

The other major plank for the WCRP is the World Ocean Circulation Experiment (WOCE). This billion-pound programme of research involving 40 countries began in early 1990. Its intensive five-year period of measurements is designed to contribute to the establishment of a scientific basis for climate prediction on time scales from years to decades. The major elements of WOCE include using oceanographic satellite observing systems (eg., TOPEX and ERS 1) to provide measurements of surface topography, surface wind and wind stress, global hydrographic, and chemical surveys; measurements of large-scale features of surface and subsurface velocities using such systems as moored current meter arrays and subsurface drifting floats; and ocean modelling studies. The central pillar of WOCE is, however, the satellite programmes. Without them the work would not be possible. The importance of these programmes is the way in which they draw together a huge range of meteorological and oceanographic work planned for the next 10 years. While most of the work would probably go ahead anyway, the provision of a coordinating thread should ensure that the sum is greater than the parts.

14.6   Conclusion

The huge range of plans for the development of almost every facet of satellite meteorology show that the subject will grow dramatically in the coming years, but it will not be plain sailing. Many of the more ambitious plans will be the subject of more and more severe financial scrutiny as costs rise. Increasingly, scientists will be required to show in advance that the satellite systems they wish to deploy can generate real economic benefits. This will place a greater pressure on the importance of extracting practical results from more advanced satellites. While meteorological applications have been among the most successful areas of satellite remote sensing, they will not be exempt from such pressures, especially where new satellites will combine both weather and other environmental observations. So the need to take a realistic view of how new systems can contribute to all forms of forecasting will be essential.

In spite of this cautionary word, there is no doubt that satellite meteorology will play an ever more important part in our lives. Thus far the benefits of weather satellites have far exceeded their costs, and there is every expectation that this success will continue. On every time scale, from improved local forecasts of rainfall a few hours ahead, to predicting possible changes in the global climate in the decades ahead, improved satellite observations will be essential if progress is to be made.

Acronyms

ATS 1	Applications Technology Satellite 1 – a geostationary satellite launched by NASA in December 1966
AVHRR	Advanced Very High Resolution Radiometer
CNES	French Centre Nationale d'Etudes Spatiales
CZCS	Coastal Zone Colour Scanner – flown on Nimbus 7
ENSO	El Nino Southern Oscillation
ERBS	Earth Radiation Budget Satellite – launched in November 1984
ERS 1	European Research Satellite 1 – due to be launched in 1991
ESA	European Space Agency
ESMR	Electrically Scanning Microwave Radiometer – flown on Nimbus 5 and 6
ESSA	United States Environmental Science Services Agency – forerunner of NOAA
FGGE	First GARP Global Experiment
GARP	Global Atmospheric Research Programme
GOES	Geostationary Observational and Environmental Satellite
HIRS	High Resolution Sounder – on the TIROS-N/NOAA satellites
IRIS	Infrared Interferometer Spectrometer
ITCZ	Intertropical Convergence Zone
ITOS	Improved TIROS Observational System
MSU	Microwave Sounder Unit – on the TIROS-N/NOAA satellites
NASA	United States National Aeronautics and Space Administration
NOAA	United States National Oceanic and Atmospheric Administration
NROSS	United States Navy Remote Ocean Sensing System
NSCAT	NASA Scatterometer
SAR	Synthetic Aperture Radar
SBUV	Solar Backscatter Ultraviolet Spectrometer
SIRS	Satellite Infrared Spectrometer – A and B instruments flew on Nimbus 3 and 4
SMMR	Scanning Multichannel Microwave Radiometer
TIROS	Television Infrared Observational Satellite

Acronyms

TOGA	Tropical Oceans and Global Atmospheres Programme
TOMS	Total Ozone Monitoring System
TOPEX	Ocean Topographic Experiment
TOS	TIROS Observational System
TOVS	TIROS Operational Vertical Sounder
VAS	VISSR Atmospheric Sounder
VISSR	Visible/Infrared Spin–Scan Radiometer
WCRP	World Climate Research Programme
WMO	World Meteorological Organisation – headquarters in Geneva, Switzerland
WOCE	World Ocean Circulation Experiment

Glossary

absolute temperature (K): a temperature scale based on the thermodynamic principle that the lowest possible temperature is absolute zero (0 K) and the ice point is 273.16 K measured in degrees Kelvin (K) which are of the same magnitude as degrees Celsius (°C).

albedo: the proportion of the radiation falling upon a nonluminous body which it diffusely reflects.

anticyclone: a region where the surface atmospheric pressure is high relative to its surroundings – often called a 'high'.

black body radiation: the radiation that is emitted by a surface which absorbs all incident radiation at all wavelengths. The wavelength-dependence of this radiation is defined by the temperature of the surface.

blocking anticyclone: a quasistationary area of high pressure which is often associated with periods of abnormal weather in mid-latitudes in the Northern Hemisphere.

cold front: the boundary line between advancing cold air and a mass of warm air under which the cold air pushes like a wedge.

coriolis force: the term used to explain the fact that a moving object detached from the rotating Earth appears to an observer on Earth to be deflected by a force acting at right angles to the direction of motion. Deflection of moving objects in the Northern Hemisphere is to the right of the path of motion. Deflection in the Southern Hemisphere is to the left of the path of motion.

cyclone: generally a name given to a region of low pressure. In temperate latitudes cyclones are usually spoken of as depressions and the term cyclone is taken to refer only to a 'tropical cyclone'.

depression: a part of the atmosphere where the surface pressure is lower than in surrounding parts – often called a 'low'.

dielectric constant: the ratio of the electric flux density produced in a material to value in free space produced by the same electric field strength.

electromagnetic radiation: all forms of electric and magnetic fields which need no medium to support them and which travel through a vacuum at the velocity of light. This radiation encompasses the entire frequency range from γ-rays to radiowaves.

El Nino Southern Oscillation (ENSO): a quasiperiodic occurrence when large-scale abnormal pressure and sea surface temperature patterns become established across the tropical Pacific every few years.

false-colour images: computer-generated images which use a range of colours to display either measurements at different wavelengths or different signal levels to highlight certain features of the observations in a visually arresting manner.

geoid: a definition of the Earth's form in terms of its gravity field which most closely approximates to mean sea level.

geostationary orbit: an equatorial orbit with a period of 24 h so that the satellite maintains the same longitudinal position over the equator at an altitude of 35 900 km.

Hadley cell: the basic vertical circulation pattern in the tropics where moist warm air rises near the equator and spreads out north and south and descends at around 20–30° N and S.

hertz (Hz): the unit of measurement of frequency. 1 Hz equals 1 cycle per second (cps).

hurricane: the name given primarily to tropical cyclones in the West Indies and Gulf of Mexico.

jet stream: strong winds in the upper troposphere whose course is related to the major weather systems in the lower atmosphere and which tend to define the movement of these systems.

katabatic wind: a wind created when very cold air forms in upland areas and becomes sufficiently dense to drain downhill.

lapse rate: a measure of the temperature profile of the atmosphere with height – the fall of temperature in a unit height.

latent heat of vaporisation: the amount of heat absorbed or emitted during the change of state from liquid to vapour or vice versa.

lidar: a technique identical to radar which uses the radiation from a visible or infrared laser to measure the distance of remote objects.

micrometre (μm): 10^{-6} metres.

millibar (mb): a unit of atmospheric pressure corresponding to a force of 1000 dynes per cm^2. Mean sea-level pressure = 1013.2 mb.

molecular spectrum: the electromagnetic absorption and emission features for each molecular species which are the unique product of its molecular structure, and which define the radiative properties of a given molecular species in the atmosphere.

orographic rainfall: the increase in rainfall that occurs when moist air is forced to rise as it passes over upland areas.

ozone: a form of oxygen (O_3) formed by photochemical action in the upper atmosphere.

photosynthesis: the process by which plants convert carbon dioxide (CO_2) and water (H_2O) of the air into carbohydrates by exposure to light.

radar: the use of radio waves or microwaves to measure the distance of objects by measuring the time taken for a pulse of radiation to travel from a transmitter to the object and back to an adjacent receiver.

radiometer: an instrument which makes quantitative measurements of the amount of electromagnetic radiation falling on it in a specified wavelength interval.

radio-sonde: a free balloon carrying instruments which transmit measurements of temperature, pressure, and humidity to ground by radiotelegraphy as it rises through the atmosphere.

signal-to-noise ratio: the ratio of the signal generated by a detector measuring radiation to the random noise generated by the detector.

spatial resolution: the lower limit to which the distance between features in a satellite image can be resolved.

spectrometer: an instrument which splits up incident electromagnetic radiation and measures the amount present as a function of wavelength.

spectroscopy: the branch of science which uses various instruments to produce and examine spectra.

Stevenson shelter: a standard housing for ground-level meteorological instruments designed to ensure that reliable shade temperatures are measured.

stratosphere: the portion of the atmosphere, typically between an altitude of 12 and 40 km, where the temperature is approximately constant and there is little or no vertical mixing motion.

Sun-synchronous orbit: a near-polar orbit in which a satellite passes over a given point of the Earth at the same time each day throughout the year.

tropopause: the boundary between the troposphere and the stratosphere.

troposphere: the portion of the atmosphere from the ground surface to around 12 km in which temperature falls with increasing height.

typhoon: a name of Chinese origin (meaning 'great wind') applied to tropical cyclones which occur in the western Pacific Ocean. They are essentially the same as hurricanes in the Atlantic and cyclones in the Bay of Bengal.

warm front: the boundary line between advancing warm air and a mass of colder air over which it rises.

warm sector: in the early stages of the life of many depressions in temperate latitudes, there is a sector of warm air, which disappears as the system deepens and the cold front catches up the warm front.

wavenumber (cm^{-1}): a measure of the frequency of electromagnetic radiation in terms of the number of wavelengths in a centimetre, where $1 \, cm^{-1} = 3 \times 10^{10}$ Hz or 30 GHz.

Annotated bibliography

THIS BIBLIOGRAPHY is designed to enable the reader to delve further into the various aspects of meteorology and satellite technology presented in this book. They provide more extensive reviews of issues covered here and the physical principles of the technologies involved, and in most cases contain extensive bibliographies of their own.

Meteorology and climatology

Barrett, E. C. (1974). *Climatology from Satellites*. Methuen: London.
    An early work which provides a useful introduction to the initial application of satellite measurements.
Barry, R. G. & Chorley, R. J. (1988). *Atmosphere, Weather and Climate*, Routledge: London.
    The fifth edition of a well-established widely read standard work which provides an up-to-date treatment of current meteorological and climatological knowledge.
Browning, K. A. (ed.) (1981). *Nowcasting*. Academic: New York.
    A series of papers which provides a guide to developments in making short-term weather forecasts using a combination of surface observation, radar systems, and satellite measurements.
Critchfield, H. J. (1983). *General Climatology*. Prentice Hall: Englewood Cliffs, New Jersey.
    The fourth edition of a popular textbook which covers the standard meteorological and climatological subjects and also their implications for various aspects of social and economic issues.
Lockwood, J. G. (1974). *World Climatology: an Environmental Approach*. Edward Arnold: London.
    A lucid and thorough introduction to the Earth's climate and its various components.
Musk, L. F. (1988). *Weather Systems*. Cambridge University Press: Cambridge.
    A comprehensive textbook which provides an introduction to meteorology and detailed information on weather systems.

Washington, W. M. & Parkinson, C. L. (1986). *An Introduction to Three-Dimensional Climate Modeling*. Oxford University Press: Oxford.
> An extensive guide to the development and use of computer models of the Earth's climate, which covers both the basic theory and the relevant numerical techniques for simulating the atmosphere, oceans, and sea ice.

Satellite technology and remote sensing

Allan, T. D. (ed.) (1983). *Satellite Microwave Remote Sensing*. Ellis Horwood: Chichester.
> A set of papers which reviews both the theoretical aspects of microwave remote sensing and the practical results of using satellite systems.

Fotheringham, R. R. (1979). *The Earth's Atmosphere Viewed from Space*. University of Dundee: Dundee, Scotland.
> A short book which provides a good selection of satellite images with a basic discussion of meteorological processes which can be observed.

Henderson-Sellers, A. (ed.) (1984). *Satellite Sensing of a Cloudy Atmosphere: Observing the Third Planet*. Taylor & Francis: London.
> A series of papers which considers the problems of remote sensing of the Earth with particular attention to measuring the atmosphere and the oceans, and the problems posed by clouds.

Houghton, J. T., Taylor, F. W. & Rodgers, C. D. (1984). *Remote Sensing of Atmospheres*. Cambridge University Press: Cambridge.
> A standard textbook which provides a clear and comprehensive description of the principles and equipment used to make remote sensing measurements of the atmosphere of the Earth and other planets.

Milford, J. R. & Reynolds, R. *Satpack 1 and Satpack 2*. Royal Meteorological Society: Reading.
> A pair of folders providing illuminating examples of satellite images together with overlay diagrams and instructions to provide classroom material to enable students to interpret these images in terms of surface data and upper air measurements.

Schnapf, A. (ed.) (1985). *Monitoring Earth's Ocean, Land and Atmosphere from Space: Sensors, Systems and Applications*. American Institute of Aeronautics and Astronautics: New York.
> An extremely wide-ranging and thorough set of papers providing information on all aspects of satellite remote sensing technology and its applications.

Scorer, R. R. (1986). *Cloud Identification by Satellite*. Ellis Horwood: Chichester.
> An extraordinary varied set of satellite images providing examples of many types of the meteorological phenomena that can be observed from space.

Climatic change

Clark, W. C. (ed.) (1982). *Carbon Dioxide Review*. Oxford University Press: Oxford.
> A series of papers and critical analyses by a wide range of authorities which thoroughly explores the many complexities of the debate about the effect of increasing carbon dioxide in the atmosphere on the Earth's climate.

Gregory, S. (ed.) (1988). *Recent Climatic Change*. Belhaven Press: London.
> A series of papers which reviews the latest evidence of climatic change in various parts of the world.

Annotated bibliography

Lamb, H. H. (1972, 1977). *Climate, Present, Past and Future*, vols. 1 and 2. Methuen: London.
   The classic work on all aspects of climatic change which considers all aspects of meteorology, climatology, the evidence of climatic change, and possible explanations of observed changes.
   (1988). *Weather, Climate and Human Affairs*. Methuen: London.
   A collection of papers and essays which complements Professor Lamb's major work on climatic change cited above.
MIT (1971). *Report of the Study of Man's Impact on Climate (SMIC)*. The Massachusetts Institute of Technology: Cambridge, Mass.
   A seminal study by many of the world's leading atmospheric scientists of the potential impact of man's activities on the climate. As such it provides the clearest possible statement of both the issues involved and the questions that need to be addressed.

Index

Index

QC
879.5
.B87
1990

QC
879.5
.B87

1990

$24.95